Mechanisms and Machine Science

Volume 183

This book series establishes a well-defined forum for monographs, edited Books, and proceedings on mechanical engineering with particular emphasis on MMS (Mechanism and Machine Science). The final goal is the publication of research that shows the development of mechanical engineering and particularly MMS in all technical aspects, even in very recent assessments. Published works share an approach by which technical details and formulation are discussed, and discuss modern formalisms with the aim to circulate research and technical achievements for use in professional, research, academic, and teaching activities.

This technical approach is an essential characteristic of the series. By discussing technical details and formulations in terms of modern formalisms, the possibility is created not only to show technical developments but also to explain achievements for technical teaching and research activity today and for the future.

The book series is intended to collect technical views on developments of the broad field of MMS in a unique frame that can be seen in its totality as an Encyclopaedia of MMS but with the additional purpose of archiving and teaching MMS achievements. Therefore, the book series will be of use not only for researchers and teachers in Mechanical Engineering but also for professionals and students for their formation and future work.

The series is promoted under the auspices of International Federation for the Promotion of Mechanism and Machine Science (IFToMM).

Prospective authors and editors can contact Mr. Pierpaolo Riva (publishing editor, Springer) at: pierpaolo.riva@springer.com

Indexed by SCOPUS and Google Scholar.

Daniel Martins · Estevan Hideki Murai

Mechanisms—Synthesis and Analysis with Applications in Robotics

Springer

Daniel Martins
Department of Mechanical Engineering
Federal University of Santa Catarina
Florianópolis, Santa Catarina, Brazil

Estevan Hideki Murai
Department of Design and Graphical Expression
Federal University of Santa Catarina
Florianópolis, Santa Catarina, Brazil

ISSN 2211-0984 ISSN 2211-0992 (electronic)
Mechanisms and Machine Science
ISBN 978-3-031-95186-2 ISBN 978-3-031-95187-9 (eBook)
https://doi.org/10.1007/978-3-031-95187-9

This Springer imprint is published by the registered company Springer Nature Switzerland AG
The registered company address is: Gewerbestrasse 11, 6330 Cham, Switzerland

Preface

The study of mechanisms is essentially a way to formalize mechanical technology. Its origins dates back almost as far as human activity itself. A common misunderstanding is that science, seen as a foundational skill, must come first in order to develop technology. In reality, technology emerged on its own, driven by humanity's need for tools. Early technologies, such as levers, were created intuitively and passed down through generations. At this point, the reader might question why we need to study the science behind these mechanisms if they can be developed intuitively. However, theory offers more than just a formal framework for technology; it serves as a valuable tool for developing technology in a more systematic and intentional way. For instance, the great technological leap that occurred in the Industrial Revolution would not have been possible without a structured approach to designing machines and mechanisms.

Mechanisms are the basis of the technology that led us to the first great Industrial Revolution. Even in the fourth Industrial Revolution, mechanisms continue to play a crucial role. Machines and robots are an important part in industries, as well as mechanisms for power transmission, payload manipulation and transportation.

In this book, we will deal with the theory of mechanisms from a geometric point of view. We believe that the geometric approach is more didactic and brings the theory of mechanisms and its applications closer to students and researchers.

In general, in Mechanical Engineering courses, the subject of mechanisms and similar subjects introduce students to the area of Mechanical Design. Therefore, it is important that since the beginning the students feel they have the ability to design and they are encouraged to develop their mechanical design skills. From this point of view, it is essential that theoretical content is presented in a more accessible way, providing the basis for design tools. Thus, in this book we show that by using a ruler and a compass you can, in a playful way, arrive at simple solutions that meet the requirements of the project.

This book is divided into three parts: Introduction to the Theory of Mechanisms, Mechanism Analysis and Mechanism Synthesis.

Part I presents the introductory concepts of mechanism theory, which will serve as the basis for analyzing and designing mechanisms. In this part, the jargon that is essential for communication among students and teacher is formed. Often students start the subject of Mechanisms with concepts that have been previously defined imprecisely or not at all formally, such as the concept of degree of freedom. The jargon and the set of concepts presented are also important for subsequent mechanical design disciplines, such as Machine Elements and Design Methodology, so we recommend that the reader pays special attention to this part of the book.

Part II presents the contents of mechanism analysis. The analysis of linkages includes important topics such as the Grashof criterion and transmission quality. The structural analysis of mechanisms presents concepts such as connectivity and redundancy, which are used in robot design. The kinematic analysis of mechanisms presents the method for carrying out the kinematic analysis of a mechanism's position, velocity and acceleration.

Part III presents tools for designing mechanisms. Among the topics covered are the number synthesis, the type synthesis and the dimensional synthesis of four-bar linkages for two and three positions. Dimensional synthesis methods for function generation and for slider-crank mechanism are also presented.

Throughout the book, there are didactic examples as well as recommended bibliographies for each topic. At the end of each chapter, there is a list of exercises of gradually increasing difficulty, designed to aid in understandings and fixing the content presented.

This book has been structured with two approaches, which are presented in this order: analysis and synthesis. However, depending on the time available to teach the subject and at the teacher's discretion, the order can be changed so that an analytical part is presented first, followed by simpler mechanism projects that require the use of past analytical theory.

It is important to note that there are a number of approaches or schools of the theory of mechanisms or, as it is currently called, the science of mechanisms and machines. The school initially led by France and later by the United Kingdom and the United States is more analytical and focuses more on the equations of dynamic and kinematic systems than on the geometric interpretation of mechanisms. The other school, which began in the second part of the 19th century, is Germanic and has a more geometric focus. It was followed by other countries, such as Eastern Europe and Russia. Each approach has its positive and negative characteristics, and in this book we have tried to integrate the concepts of these schools in order to provide a valuable set of tools for the analysis and design of mechanisms.

Finally, this publication is the result of the work experience, study and teaching of the theory of mechanisms of the first author since 1990 and of the second author since 2010. We would like to thank everyone who has contributed in any way to this book and wish you a pleasant reading.

Florianópolis, Brazil
October 2025

Daniel Martins
Estevan Hideki Murai

Contents

Part I
Introduction to the Theory of Mechanisms

Chapter 1
Introduction

Abstract This chapter establishes the foundational relevance of mechanisms and machines through a historical and societal lens. It emphasizes how ancient civilizations utilized mechanical innovations to build infrastructure, wage war, and expand territories. Examples include the Egyptian shadoof for irrigation, Chinese south-pointing chariots for navigation, and Archimedean devices in Greece for engineering and warfare. The chapter invites the reader to appreciate the cultural, functional, and technological contexts in which mechanisms evolved, setting the tone for deeper technical exploration. The chapter also presents basic concepts on Mechanism Science, establishing definitions which are necessrary for the development of further topics on mechanism synthesis and analysis. It serves as a conceptual entry point into the discipline.

Keywords Basic concepts · Mechanisms components · Kinematic chain types · Kinematic pairs types · Mechanism representations

The study of machines and mechanisms is basically the study of history of societies. The construction of machines and mechanisms can affect the course of civilizations. At first glance, this statement may seem a bit odd or dramatic; however, it is true.

In ancient Egypt, rustic but ingenious machines and mechanisms were created to cut rocks, move them around, and as tools in the construction of houses and temples. Apart from their conventional use, such works (pharaonic in their strict sense) demonstrated the power of the empire. Figure 1.1a shows a *shadoof*, a device developed in ancient Egypt to assist in the irrigation of crops [1].

At the same time in China, many mechanisms already existed to serve society and also the army [2]. There are records that in 132 AD the first seismographs appeared. Other Chinese developments are water-powered clocks (around 1088 AD), carts that always pointed south when being driven (used mainly for logistics in Chinese armies, around 200 AD), mechanical animals such as horses and oxen (used mainly for entertainment, around 480 BC) and also automata for serving tea, writing and impressing visitors. Figure 1.1b shows a reconstruction of the south-pointing chariot developed in ancient China.

D. Martins and E. H. Murai, *Mechanisms–Synthesis and Analysis with Applications in Robotics*, Mechanisms and Machine Science 183,
https://doi.org/10.1007/978-3-031-95187-9_1

* (a) * (b)

Fig. 1.1 Ancient mechanisms. **a** *shadoof* used for irrigation. **b** south-pointing chariot. (Adapted from [4])

In ancient Greece, Archimedes developed a number of mechanisms that resulted in competitive advantages of the Greek civilization over other contemporary civilizations, such as the Archimedes' screw (see Fig. 1.2), the Archimedes' claw, the lever principle, and improvements in the power and accuracy of catapults [3].

Siege weapons are examples of mechanisms and machines that changed human history. Examples of siege weapons include catapults, trebuchets (see Fig. 1.3), mangonels (or traction trebuchets), ballistas, onagers, siege towers, and battering rams. Such weapons provided the warlike advantage used to create, expand, and defend empires.

Fig. 1.2 Archimedes' screw used for water lifting

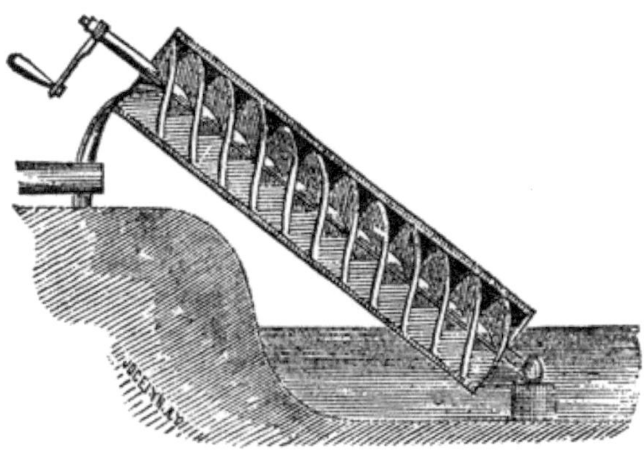

Fig. 1.3 Trebuchet used to throw heavy projectiles

Navigation and positioning mechanisms aided in the exploration, mapping and territorial expansion of civilizations as well as the colonization of new territories. Examples include the south-pointing chariot, the Antikythera machine, the gyroscope, and some more sophisticated versions of the astrolabe [2, 5].

In civil society, mechanisms have also had great influence. The dissemination of culture was accelerated after the invention of the printing press by Johannes Gutenberg in the 15th century (see Fig. 1.4a). On the other hand, the invention of the typewriter made writing faster and more objective.

The big difference between today's society and medieval society is largely due to the Industrial Revolution. This revolution marked the shift from manual production to mass production, and behind this revolution were machines. Steam-powered machines generated mechanical power for an entire industry. Figure 1.4b shows a drawing of Watt's steam engine. Figure 1.5 shows the courtyard of a 19th century factory. Such a factory had a central steam engine whose power was transmitted to all machines in the factory through belts.

The sewing machine is an example of a mechanism that, despite being simple, has generated great changes in society. In the 19th century, before the sewing machine, mending and sewing new clothes were tasks that occupied a large part of women's day. With the popularization of the sewing machine, the hours previously spent sewing became free, contributing to the insertion of women in the labor market. A curious aspect of the sewing machine is that this machine is arguably the first complex machine that entered common people houses. Later, the invention of the washing

*

(a)

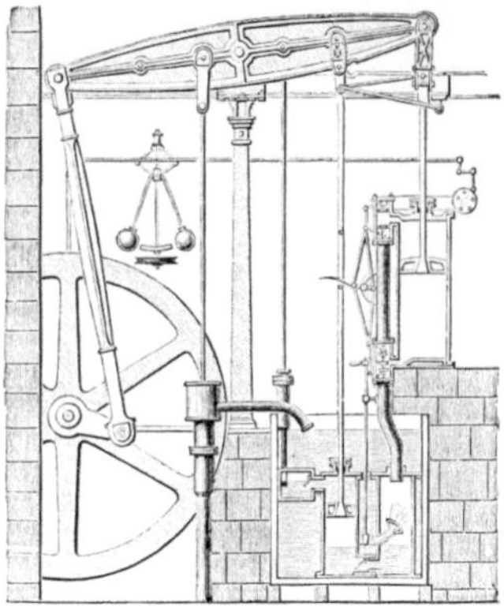

*

(b)

Fig. 1.4 Mechanisms that influenced civil society. **a** Gutenberg's printing press. **b** Watt's steam engine

Fig. 1.5 Nineteenth century factory powered by a steam engine

machine had a similar effect, making washing more time efficient and increasing the profits of women who washed clothes for others or even for their own family.

In transportation, the invention of the steam locomotive, the automobile, the airplane and motor boats catalyzed the development of society.

The mechanization of agriculture resulted in the production of foods on an industrial scale, making it possible to feed societies with large populations.

In these brief examples of machines that changed society, the importance of machine design becomes evident. Usually, the machine design starts with the kinematic design and then goes on to the structural dimensioning and specification of components. This book will cover the kinematic design of machines.

Finally, it is observed that there are common areas between some engineering disciplines, such as Thermal Sciences (also studied by chemical engineers) and Materials Science (also studied by materials engineers); however, the design and manufacture of machines and mechanisms are studied only in Mechanical Engineering and its derived engineering disciplines (naval, automotive, aerospace, among others).

1.1 Basic Concepts

The following sections present basic concepts and definitions of mechanism theory. These concepts are fundamental for the development and correct understanding of the next chapters. For more information regarding the definitions presented here, see [6–13].

1.1.1 Kinematics

Kinematics is the branch of theoretical physics, specifically mechanics, that studies the motion of particles and bodies with no regard to the actions that generated such motions.[1] Studying the behavior of a pendulum in simple harmonic motion (SHM) and determining the time until a clash between two cars on a road are typical kinematic problems (see Fig. 1.6).

Applied kinematics studies kinematics in the scope of engineering problems, such as the design and analysis of machines and mechanisms. The study of kinematics applied to machine design is also known as *Mechanism and Machine Theory* (MMT) or, more modernly, *Mechanism and Machine Science* (MMS). If it is necessary to consider the actions involved and not just the motions, one can do a static or dynamic analysis.

For example, in mechanisms one often studies gear kinematics in order to specify gears so that they meet design requirements. Another example of applied kinematics is the study of cams, aiming to determine the cam geometry in order to obtain the desired motion in the follower. Yet another example of applied kinematics is the study of linkages in order to determine the dimensions required for the mechanism to do the desired task. The study of the kinematics of gears, cams, and linkages are typical problems of applied kinematics for the study of mechanisms (see Fig. 1.7).

[1] The term *action* can refer to both forces and moments.

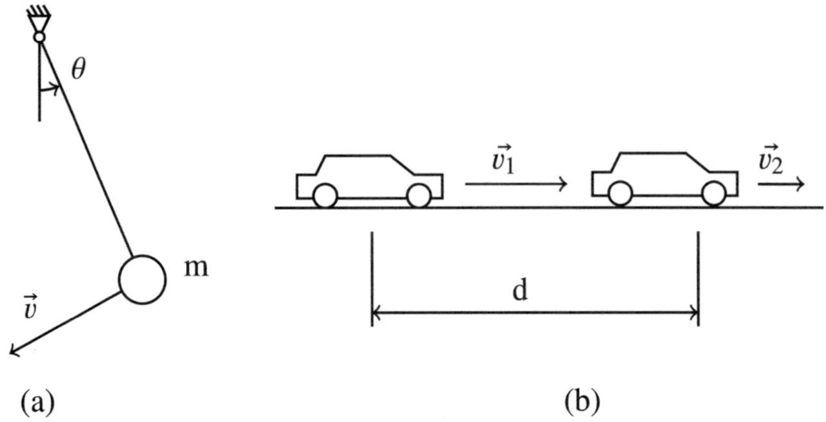

Fig. 1.6 Typical examples of kinematics. **a** Pendulum in SHM. **b** cars in motion

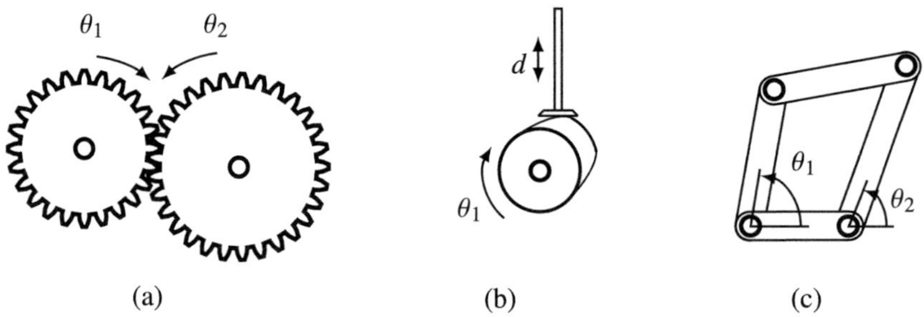

Fig. 1.7 Typical examples of kinematics applied to the study of mechanisms. **a** gears, **b** cam, **c** linkage

1.1.2 Particles, Rigid Bodies and Mechanical System

Particle is a mechanical system with negligible dimensions but with an internal structure to which a physical quantity, usually scalar such as mass, can be associated. As examples, one can cite the size of a car along a road and the diameter of a planet in relation to its orbit.

Premise of rigidity is the condition satisfied by particles that hold the distances between them constant in time. Thus, let there be any two particles, A and B, and the distance between them given by $d(A, B)$. The two particles satisfy the rigidity premise if

$$\frac{\partial d(A, B)}{\partial t} = 0. \tag{1.1}$$

Body is a continuous collection of particles. A **rigid body** is a theoretical model of a solid body whose particles satisfy the principle of rigidity, that is, the body will not deform even if a force is applied to it. Note that it is not necessary that a rigid body

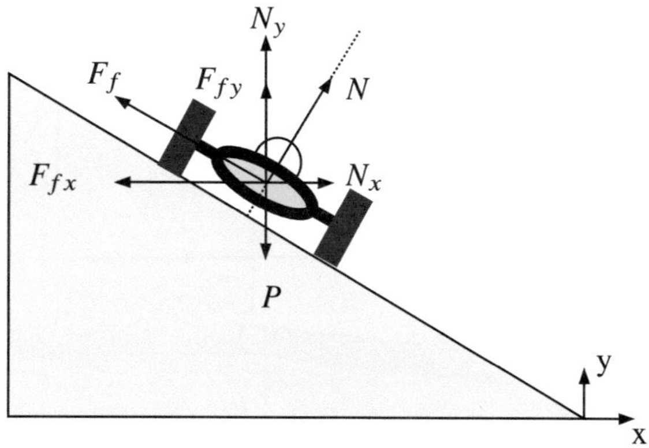

Fig. 1.8 Mechanical system consisting of a car, the track, and the interaction forces

be isotropic or homogeneous, so reinforced concrete or composite materials could be considered as rigid bodies in some cases. On the other hand, a **deformable body** is a continuous set of particles (body) that does not satisfy the premise of rigidity.

Mechanical system is a set of particles and/or rigid bodies and their interrelationships. Such interrelationships are given through forces, either field or contact forces. An example of a mechanical system is shown in Fig. 1.8, in which a car interacts with the Earth through three forces, the gravitational force (field force), the normal force (contact force) and the frictional force (contact force). In Fig. 1.9 another example of a mechanical system is presented. In this case, the suspension mechanism of the car from Fig. 1.8 is being analyzed.[2]

1.1.3 Reference System

A **reference system** or referential is a coordinate system fixed at a point used to measure and characterize physical quantities such as position, velocity, and acceleration. Often a basis of unit vectors that are perpendicular to each other and dextrous is used to associate the oriented axes of the coordinate system representing the reference system. This coordinate system is then fixed at a point. The vector basis can be defined by a set of three **non-collinear** points, P_1, P_2 and P_3, as shown in Fig. 1.10a. Thus, through P_1 and P_2 one can define the x axis of the coordinate system (see Fig. 1.10b), through P_3 and perpendicular to the x axis one can define the y axis of the coordinate system. Finally, the z-axis can be defined by the line perpendicular to the x and y axes, in order to generate a right-handed basis. Note that the relative positions of a point with respect to the other two must be time invariant, in order to obtain a unique coordinate system that is also time invariant.

[2] The suspension spring has been omitted in Fig. 1.9 for better visualization.

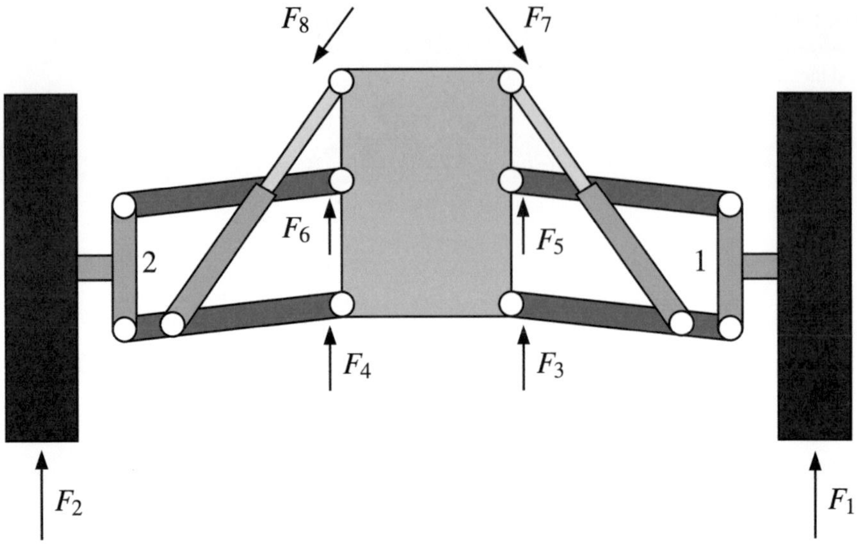

Fig. 1.9 Simplified mechanical system of a car suspension

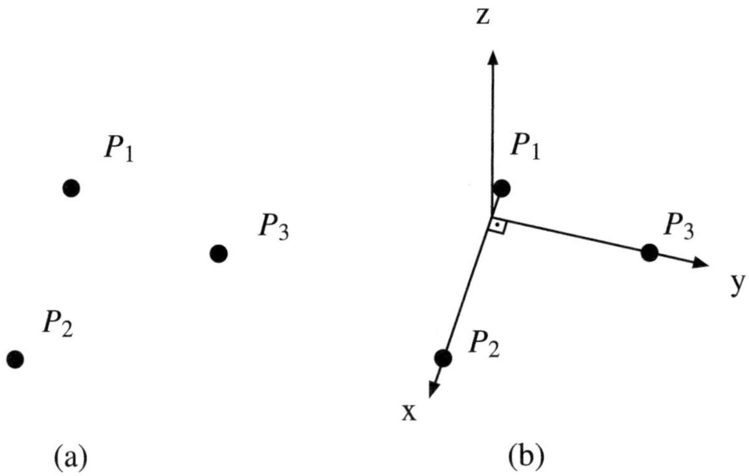

Fig. 1.10 Determining a coordinate system. **a** Three non-collinear points, **b** derived coordinate system

Absolute reference system (also called **fixed reference system**) or absolute referential is a reference system adopted to represent the entire behavior of a mechanical system. **Moving reference system** or moving referential is a reference system attached to a moving body. There are several options for choosing a moving reference system, and a mechanical system can contain as many moving reference systems as are convenient for mathematically modeling the system. On the other hand, even

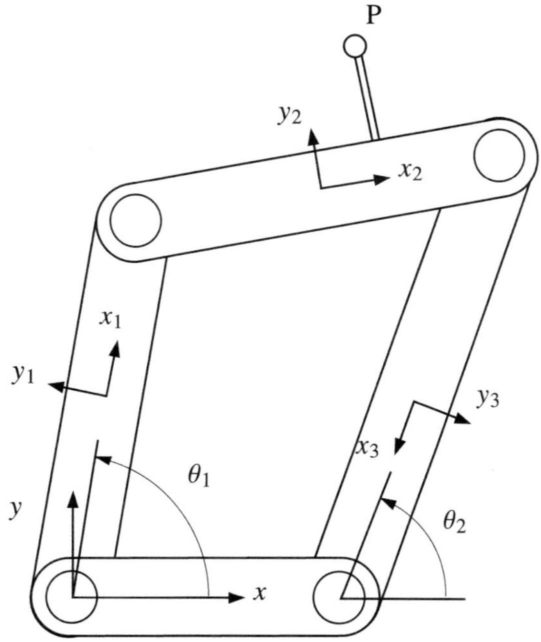

Fig. 1.11 Absolute and moving reference systems

though there are several possibilities for choosing an absolute reference system, once an absolute reference system is adopted, the mechanical system will possess only that fixed reference system.

For example, in Fig. 1.11 we have one absolute reference system, given by the xy coordinate system, and three moving reference systems, given by the x_1y_1, x_2y_2, and x_3y_3 coordinate systems. When the links move, the reference systems x_1y_1, x_2y_2, and x_3y_3 move according to the links with which these reference systems are associated. Using moving reference systems can be convenient when you want to express the position of a point that does not move with respect to a specific link. For example, if the position of the point P in Fig. 1.11 is measured directly in the absolute reference system, its coordinates P_x and P_y will vary in time. Also, if the location of P on the link is changed, P_x and P_y will change. However, if the position of P is measured with respect to the reference system x_2y_2, obtaining the coordinates P_{x_2} and P_{y_2}, to obtain the same coordinates with respect to the absolute reference system (P_x and P_y) a coordinate transformation is sufficient. Note that P_{x_2} and P_{y_2} will not change with time. When the location of P changes, simply change the coordinates P_{x_2} and P_{y_2} to obtain the coordinates P_x and P_y. Thus, when measuring the position of a point P on a link, it is more convenient to use a coordinate system associated with the link itself than coordinate systems associated with other links.

1.1.4 *Mechanisms and Their Components*

Kinematic chain is a collection of rigid bodies subject to mechanical constraints. A kinematic chain can be classified into the types below.

Open or serial: kinematic chain in which starting from a rigid body there is only a single possible path to reach another rigid body. There are at least two rigid bodies that are connected to only one rigid body. An example of an open kinematic chain is shown in Fig. 1.12a.

Closed or parallel: kinematic chain in which starting from a rigid body there are at least two possible paths to reach another rigid body. Every rigid body is connected to at least two other rigid bodies. Examples of closed kinematic chains are shown in Figs. 1.12b and 1.13.

Hybrid or mixed: kinematic chain that presents simultaneously open and closed subchains. An example of a hybrid kinematic chain is shown in Fig. 1.12c.

Comparing open kinematic chains with closed chains, their advantages (+) and disadvantages (−) are noted.

Open chains:

- need actuators in all kinematic pairs;
- have the cumulative position errors along the kinematic pairs;

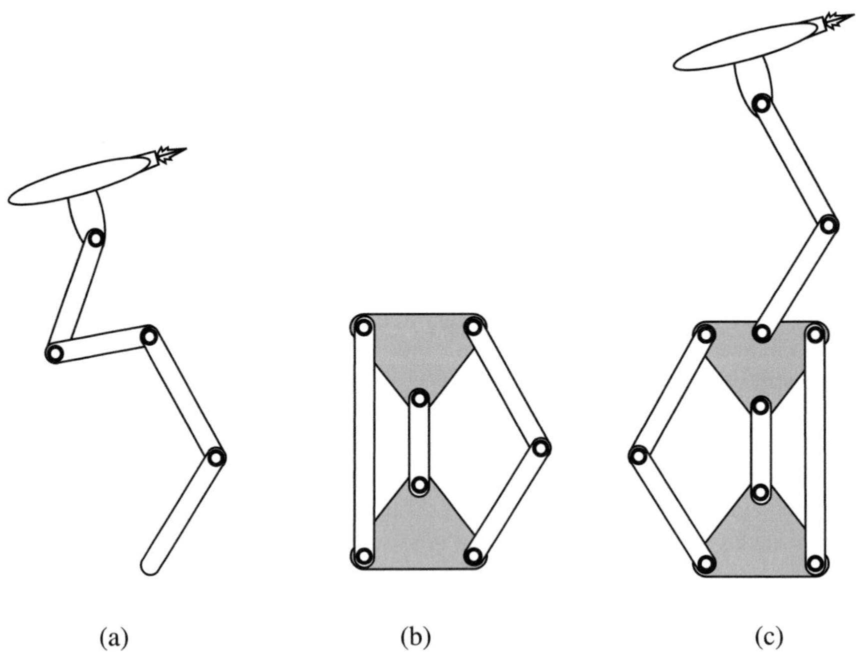

Fig. 1.12 Types of kinematic chains. **a** Open chain, **b** closed chain, **c** hybrid chain

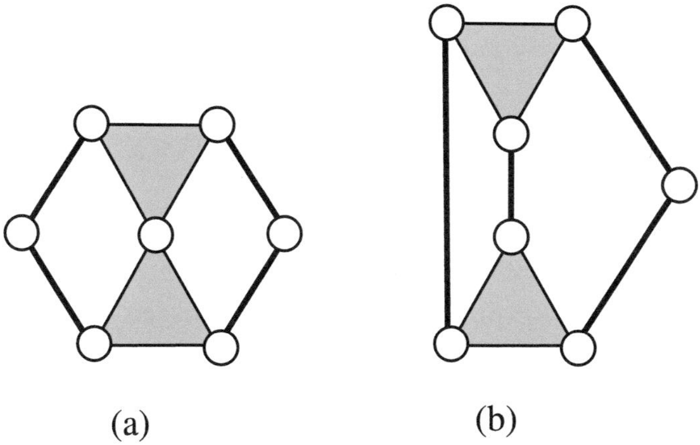

Fig. 1.13 Examples of closed kinematic chains. **a** Watt's chain, **b** Stephenson's chain

- have loads to be supported cumulative along the links, resulting in sequentially more powerful actuators and sequentially more robust and heavy links;
+ have larger working space.

Closed chains:

+ do not require actuators in all kinematic pairs. One can take advantage of this characteristic by judicious positioning the actuators. For example, if actuators are positioned on the kinematic pairs that connect the fixed link to the other links, the weight of these actuators can be supported directly by the fixed link, not adding mass on the moving links;
+ are more accurate, because the position errors are not cumulative;
+ can have the loads to be supported distributed among several links, requiring less robustness, reducing the links inertia and increasing the maximum accelerations and speeds achieved;
- present a more restricted working space. Thus, the design must be done with more caution, observing the working area that the mechanism must have and the working area that the mechanism has.

Link is a rigid body that belongs to a kinematic chain. In Fig. 1.15a, a kinematic chain with the identified links is displayed. The number of links in a kinematic chain is represented by the symbol n. You can classify links by the number of connections that each link has (see kinematic pair elements). Thus, if a link has one connection it is said to be *unary*, if it has two connections it is a *binary* link, if the link has three connections it is a *ternary* link, and so on. Links with three or more connections are called *polygonal* links.

Coupling is a connection between two links in a kinematic chain. This connection limits the possible motions that one link has in relation to the other. In this way, couplings impose mechanical constraints on the links in a kinematic chain. Couplings can be classified into the two types below.

Direct coupling: coupling in which there is direct contact between two links in a kinematic chain. Figure 1.15b shows a direct coupling between links 2 and 3 of the open kinematic chain.

Indirect coupling: coupling in which two links of a kinematic chain are not in contact. In Fig. 1.15c an indirect coupling between links 2 and 4 of the open kinematic chain is shown.

In addition to the forces generated by the physical coupling between the links there are also gravitational and electromagnetic field forces, which in some cases can be neglected.

Kinematic pairs are the direct couplings of a kinematic chain. The number of kinematic pairs in a kinematic chain is represented by the letter p. Throughout this book, kinematic pairs are also referred to simply as *pairs*, implying kinematic pairs by context. A kinematic pair can be classified into the two types below.

Lower: kinematic pair whose contact between the two links is through a surface.

Higher: kinematic pair whose contact between the two links is through a curve or point.

Figure 1.14 shows the classification of direct and indirect couplings.

Kinematic pair elements are the parts of each link that come into contact in the kinematic pair. Note that in lower pairs the kinematic pair elements are always surfaces, while in higher pairs the kinematic pair elements can be curves or points.

Lower kinematic pairs offer several advantageous characteristics when compared to higher kinematic pairs. One significant advantage lies in their distribution of forces. In lower kinematic pairs, the contact between elements occurs on a surface, enabling

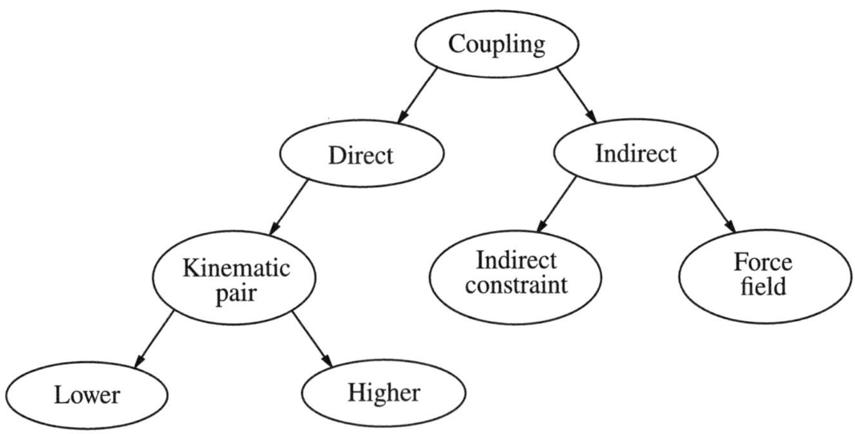

Fig. 1.14 Couplings and kinematic pairs types

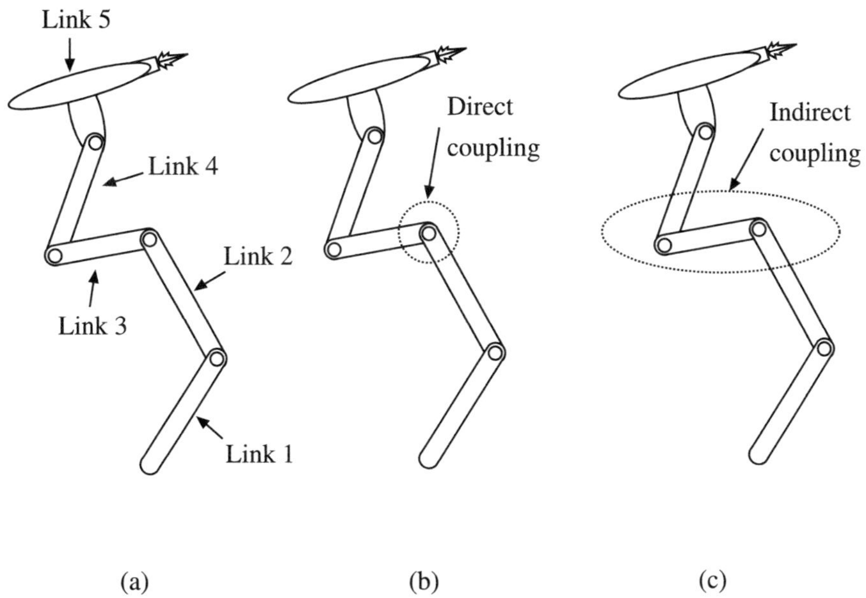

Fig. 1.15 Components of mechanisms. **a** Links of a kinematic chain, **b** direct coupling between links 2 and 3, **c** indirect coupling between links 2 and 4

the forces transmitted between links to be spread across this surface. Conversely, in higher kinematic pairs, forces are either distributed along a contact curve or concentrated at a point of contact, leading to potentially higher stresses. Consequently, lower kinematic pairs typically experience lower stresses, resulting in longer life spans, improved lubrication conditions, and reduced maintenance requirements.

According to the type of motion allowed, lower kinematic pairs can be classified into the six types below.

Revolute kinematic pair (R): kinematic pair that allows one link to rotate relative to the other link, as shown in Fig. 1.16a.

Prismatic kinematic pair (P): kinematic pair that allows one link to translate relative to the other link, as shown in Fig. 1.16b.

Helical kinematic pair (H): kinematic pair that allows one link to translate and rotate relative to the other link, with translation and rotation being coupled through pitch h. A helical kinematic pair is shown in Fig. 1.16c.

Cylindrical kinematic pair (C): kinematic pair that allows one link to translate and rotate relative to the other link, with translation and rotation being uncoupled. A cylindrical kinematic pair is shown in Fig. 1.16d.

Planar kinematic pair (E): kinematic pair that allows one link to make two translations and one rotation with respect to the other link, with the rotation axis being normal to both translations axes. The letter E is commonly adopted to represent planar kinematic pairs due to the German school of mechanism theory, in which the word *ebene* is used for planar-type kinematic pairs. An example of a planar kinematic pair is shown in Fig. 1.16e.

Spherical kinematic pair (S): kinematic pair that allows one link to make all three rotations with respect to the other link. The three axes of rotation must concur at a single point. A spherical pair is shown in Fig. 1.16f.

Examples of higher kinematic pairs are shown in Fig. 1.17. Some kinematic pairs allow more than one motion between the links. For example, the cylindrical pair allows independent rotation and translation between the two links. A kinematic pair that allows n motions can be replaced by n pairs, each allowing one motion. This replacement is called a **kinematic pair expansion**. On the other hand, one can replace a set of n kinematic pairs with a single kinematic pair that allows all the motions of the n pairs. This substitution is called a **kinematic pair contraction** . Figure 1.18 shows the expansion and contraction of a cylindrical kinematic pair.

Joint is the physical realization of the kinematic pair, and may even have intermediate components between the links (indirect coupling). Thus, while the kinematic pair refers to the motion that one link has in relation to another link (that is, it concerns the kinematics), the joint concerns the constructive type of the kinematic pair. Thus, the same kinematic pair can be realized through several different joints.

For example, a revolute pair can be realized through a ball bearing, a journal bearing, or even a pin-in-hole type joint. Each of these realizations is a joint, however, they all exhibit the same motion: a rotation. Similarly, a helical kinematic pair can be realized using a trapezoidal profile, a square profile, or recirculating ball screw.

Note that there are kinematic pairs whose type of construction uses intermediate kinematic pairs. For example, a bearing allows the rotation between the two links, being a lower kinematic pair. However, if this bearing is a ball bearing, then the kinematic pairs between the rings and the rolling elements are point contact pairs (higher kinematic pairs), presenting the disadvantages of such type of kinematic pair. The higher kinematic pairs in ball bearings are highlighted in Fig. 1.19.

An **actuator** is a device that generates a relative motion between two parts from an input signal. Examples of actuators commonly found in mechanisms are electric motors, pneumatic and hydraulic cylinders.

1.1.5 Machines and Mechanisms

Two concepts that are often confused are machine and mechanism whose definitions are given below.

Machine is a mechanical system that does a specific task, such as the transformation of a material (machining, forming, among other manufacturing processes). A machine accomplishes its task by transferring motions (rotation or translation) or actions (force or momentum) or by transforming motions into actions and vice versa.

Mechanism is a system of bodies designed to transform motions and/or actions of one body into motions and/or actions in another body. The term *mechanism* is also

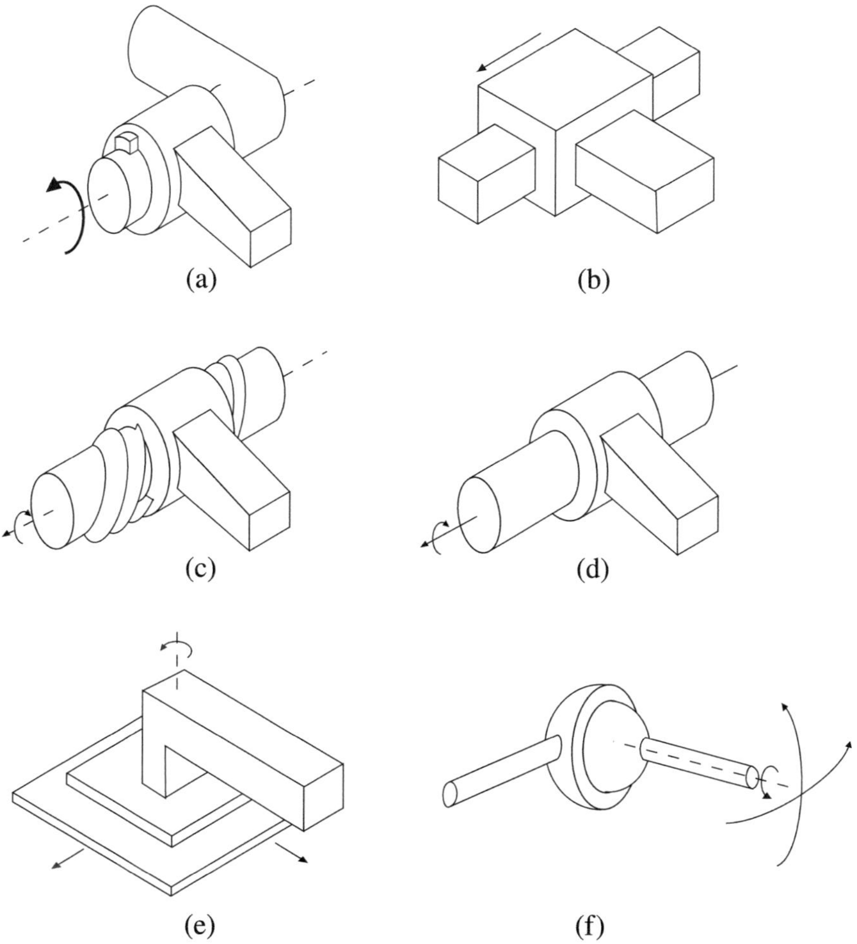

Fig. 1.16 Lower kinematic pairs. **a** revolute kinematic pair. **b** prismatic kinematic pair. **c** helical kinematic pair. **d** cylindrical kinematic pair. **e** planar kinematic pair. **f** spherical kinematic pair

used to refer to a kinematic chain in which an absolute referential has been associated with one of the links. The link to which the absolute referential has been associated is said to be the *fixed link*.

Although there is no clear line separating these two concepts we will try to give the two perspectives here. Examples of mechanisms that transform input motions into output motions are mechanisms for turning pages of books and for positioning the magnetic tape of VHS tapes. Theo Jansen's strandbeest mechanism transforms the rotation of an axis into a gait like motion (see Fig. 1.20a). When inserting the tape in VHS players, a device carries the magnetic tape to the reading head. This mechanism is driven by an electric motor and it transforms the rotation into the desired motion

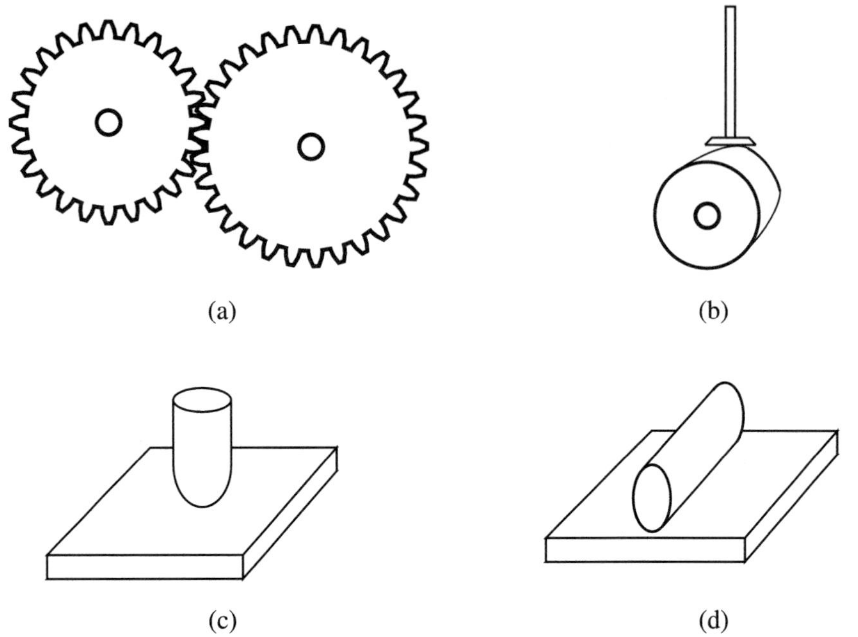

Fig. 1.17 Higher kinematic pairs. **a** Gear, **b** cam, **c** point contact, **d** linear contact

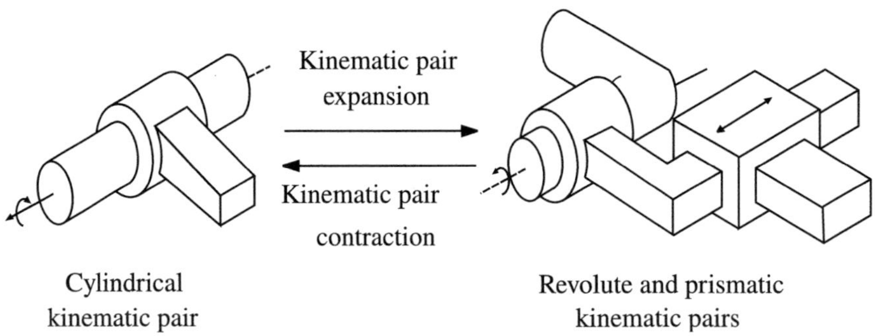

Fig. 1.18 Expansion and contraction of kinematic pairs

to position the magnetic tape, Fig. 1.20b indicates the rollers that guide the magnetic tape towards the reading head.

Examples of mechanisms that transform input actions into output actions are gearboxes and cutters. Gearboxes use gears in order to obtain an output torque greater than the input torque, as shown in Fig. 1.21a. Cutting pliers use the lever principle to increase the output force.

Finally there are mechanisms where both motion transformation and action transformation are desired, such as automotive transmissions. In transmissions, low ratio

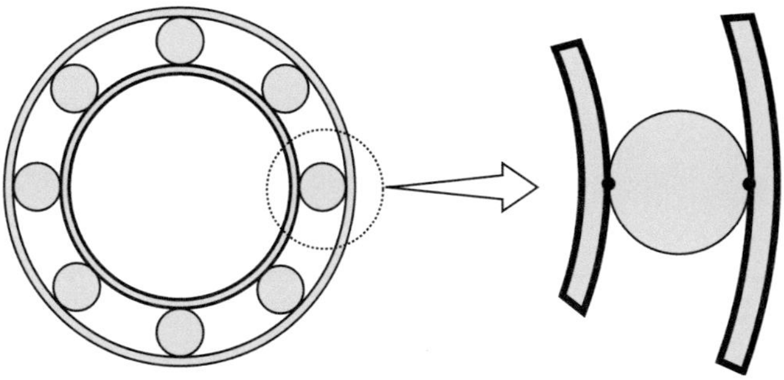

Fig. 1.19 Point contact kinematic pair in ball bearings

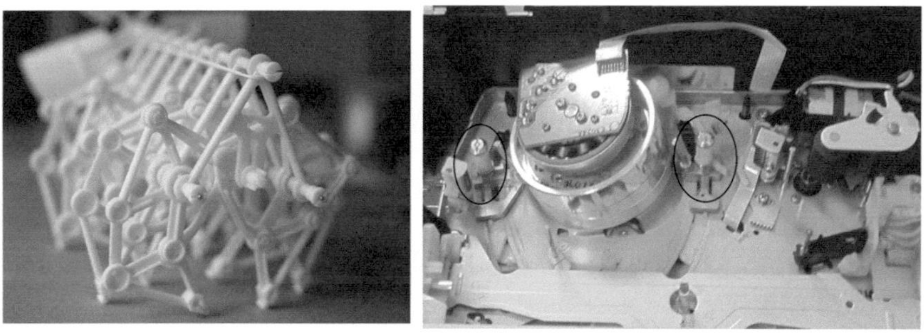

* (a) * (b)

Fig. 1.20 Mechanisms that transform motion. **a** Theo Jansen's strandbeest (adapted from [14]), **b** VHS tape loader

gears are used to achieve more torque at the output shaft. On the other hand, higher gears are used to obtain higher speeds. A transmission is shown in Fig. 1.21b.

Linkage is a mechanism where all kinematic pairs are lower pairs. Several examples of linkages are shown in Figs. 1.22, 1.23, and 1.24.

Figure 1.22a shows Theo Jansen's mechanism, used to simulate the motion of animal legs. Figure 1.22b shows the Peaucellier-Lipkin mechanism, used to make straight lines. Figure 1.22c shows the flap mechanism of a Boeing 787. Figure 1.22d shows another flap mechanism.

Figure 1.23a shows the slider-crank mechanism used in internal combustion engines. In Fig. 1.23b, a bicycle suspension mechanism is shown and in Fig. 1.23c the Honda CB 1000 suspension mechanism is shown. Figure 1.23d shows Stewart's platform, used in flight simulators.

In Fig. 1.24a a skid steer is shown. Figures 1.24b and 1.24c show garbage trucks. In Fig. 1.24d the mechanism of an excavator is shown.

* (a) * (b)

Fig. 1.21 Mechanisms whose transformation involves actions. **a** Reducer, **b** automotive transmission

1.1.6 Mechanisms Representations

A mechanism can be represented in several ways. The most common representations are design representation, functional representation, structural representation, and graph representation. A bicycle suspension in all four representations is shown in Fig. 1.25.

In the **design representation**, the mechanism components are represented as in the mechanism physical embodiment, preserving the links and joints geometry and proportion. Physical characteristics, such as the link curves and width are regarded. These are relevant characteristics for mechanism and machine design, however, they can often be negleted at the initial stages of kinematic synthesis and analysis. An example is shown in Fig. 1.25a.

In the **functional representation**, the links and kinematic pairs of the mechanism are shown in schematic form. Thus, it can be seen that the functional representation of a prismatic pair visually resembles the motion allowed by the kinematic pair. An example is shown in Fig. 1.25b.

The **structural representation** shows the types of links. Binary links are represented by lines and polygonal links by polygons with the same number of vertices as the number of kinematic pair elements the link has. The kinematic pairs connecting the links are expanded and each degree of freedom is represented by a point. It is also possible that a point represents more than one degree of freedom. In this case, the type of kinematic pair must be indicated in the drawing. An example is shown in Fig. 1.25c.

In the **graph representation**, the links are represented as vertices and the expanded kinematic pairs are represented by edges. A practical way to obtain the

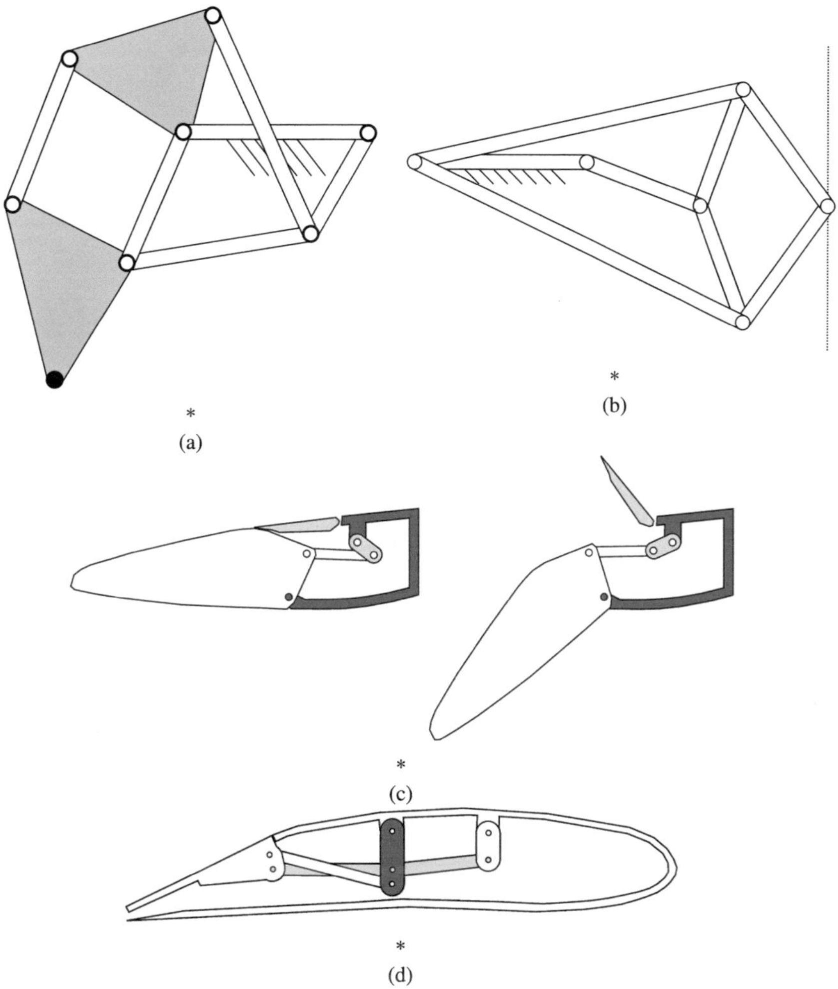

Fig. 1.22 Linkages. **a** Theo Jansen mechanism, **b** Peaucellier-Lipkin mechanism, **c** flap mechanisms of Boeing 787, **d** flap mechanism

graph representation from the structural representation is to place a vertex in the middle of each link, and then connect the vertices by edges according to the connection of links by kinematic pairs. An example is shown in Fig. 1.25d.

The four representations previously explained for mechanisms can also be used to represent kinematic chains by simply not indicating the link to which an absolute reference system has been associated. Figure 1.26 shows an abstract and fidelity scale among the four mechanism representations.

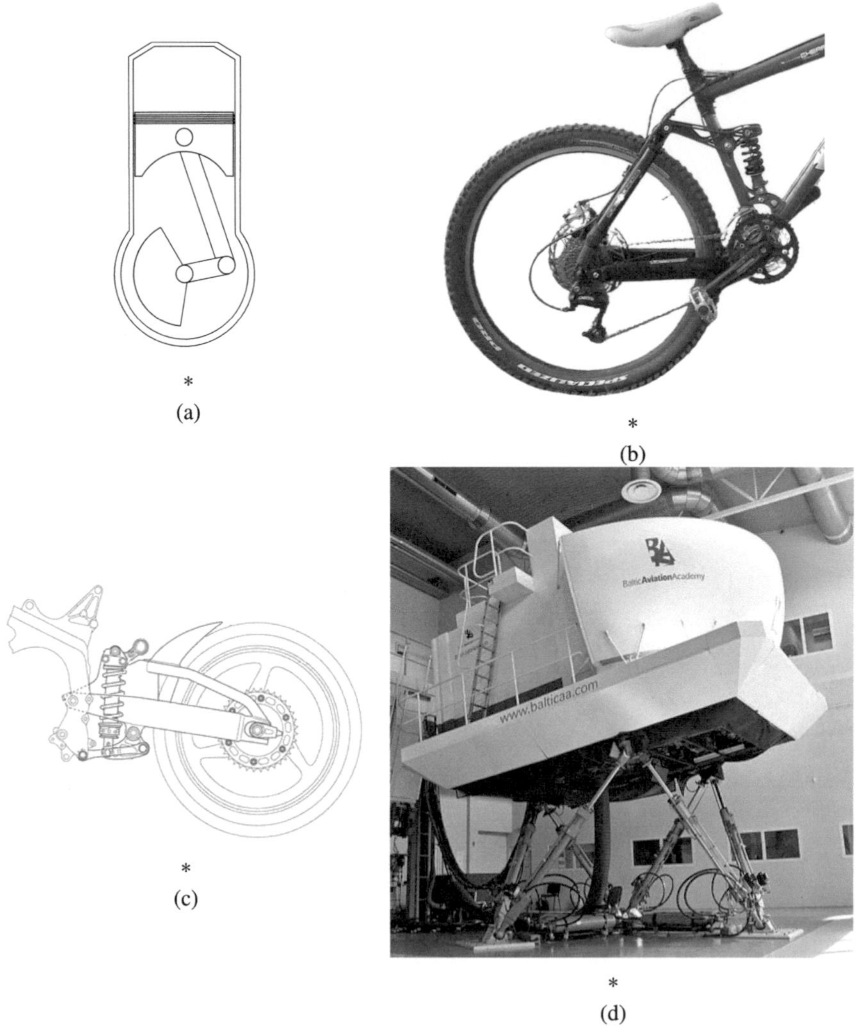

Fig. 1.23 Linkages. **a** slider-crank mechanism, **b** bicycle rear suspension mechanism, **c** Honda CB 1000 rear suspension mechanism, **d** Stewart's platform for flight simulator

1.1.7 Structural Characteristics

The **structure of a mechanism or kinematic chain** is the number and type of the links and kinematic pairs in a mechanism as well as the sequence in which the links are connected by the kinematic pairs. Thus, structural features are characteristics that refer to how the links are connected to each other, *independent of the links dimensions*. Initially, the concept of degree of freedom is introduced, then two structural

*

(a)

*

(b)

*

(c)

*

(d)

Fig. 1.24 Linkages. **a** skid steer, **b** dumpster lift mechanism on a garbage truck, **c** garbage compactor mechanism on a garbage truck, **d** excavator arm mechanism

characteristics are be presented: the workspace and the number of independent loops. More structural characteristics are presented later on this book.

Degree of freedom is the number of independent variables required to unambiguously define the configuration of a system. For example, to completely determine the positioning and orientation of a body in space requires six parameters, the position coordinates x, y and z and the orientation coordinates θ_x, θ_y and θ_z.

In mathematics, when considering functions of several variables, in order to obtain the value of the function it is necessary to know the value of each variable. Therefore, the degree of freedom of a function is the number of an dependent variables it has.

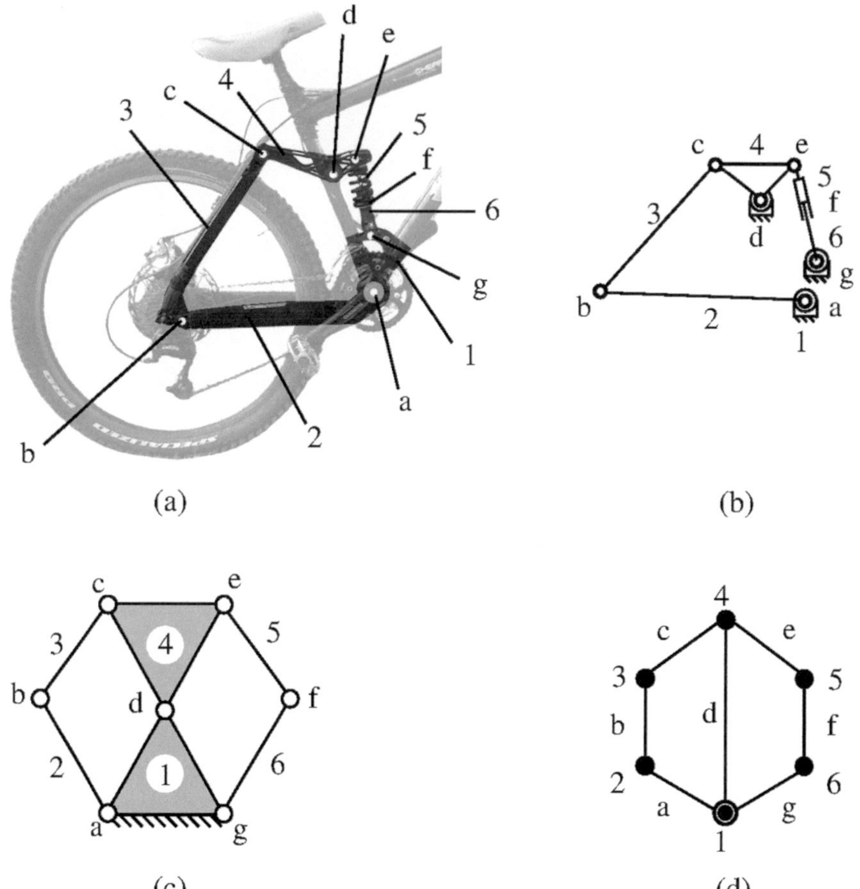

Fig. 1.25 Representations of mechanisms. **a** design representation, **b** functional representation, **c** structural representation, **d** graph representation

Workspace is the space in which the mechanism operates.[3] The **dimension of the workspace** (λ) is given by the number of independent motions possible within that space. This space may be contained in the plane, in space, or in some other subspace. In this book, we will mostly study mechanisms whose workspace is contained in the plane, and in some cases mechanisms whose workspace is contained in the space itself.

In the planar workspace, three motions are possible: the two translations contained in the plane and the rotation whose axis is normal to the plane. Thus, the dimension

[3] Mathematically speaking, the workspace is the vector space generated by the linear combination of the vectors that represent the motions allowed by each kinematic pair. These vectors that are linearly combined are the basis vector of the workspace. Thus, the dimension of the workspace is the number of basis vectors of this vector space.

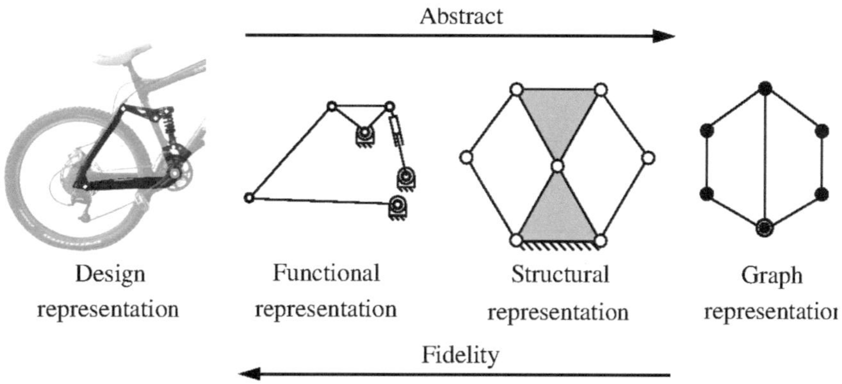

Fig. 1.26 Abstract and fidelity in mechanism representations

of this workspace is **three** ($\lambda = 3$). Mechanisms whose workspace is contained in the plane are said to be planar mechanisms, and any point on the mechanism executes a trajectory contained in a plane.

In the spatial workspace, six motions are possible: the three translations and the three rotations. Thus, the dimension of this workspace is **six** ($\lambda = 6$). Mechanisms whose workspace is contained in space are said to be spatial mechanisms.

Some examples of other workspaces are spherical space and gear space. The spherical space allows all three rotations, having dimension **three** ($\lambda = 3$). Analyzing the gear workspace is not so trivial. The kinematic pair connecting the teeth of two gears is a linear contact kinematic pair (see Fig. 1.17d). Thus, two teeth are in contact across a line. One tooth can rotate relative to the other around the contact line. Also, one tooth can translate relative to the other and still maintain a contact line by sliding one tooth over the other, translating the contact line over the tooth profile. In this way, the gear space allows one rotation and one translation, and therefore has dimension **two** ($\lambda = 2$).

Loop is a subset of links and kinematic pairs that form a closed loop. The **number of independent loops** (ν) is the number of loops that have at least one unique kinematic pair. For example, Fig. 1.27a shows a kinematic chain that has only one independent loop (ν_1), consisting of the pairs a, b, c and d. Figure 1.27b shows a kinematic chain with two independent loops ($\nu = 2$), which can be:

$$\nu_1 = \{a, b, f, g\} \text{ and } \nu_2 = \{c, d, e, f, g\}$$

Note that the pairs f and g belong to both loops, however, both ν_1 and ν_2 present unique pairs. Loop 1 presents pairs a and b as exclusive and loop 2 presents pairs c, d and e as exclusive. It is possible to assemble a third loop, for example, the outer loop $\nu_3 = a, b, c, d, e$; however, all pairs of ν_3 are already part of ν_1 or ν_2. Thus, ν_3 has no unique pair, so it is not considered an independent loop. Still, one could have chosen two other independent loops, such as $\nu_1 = a, b, f, g$ and $\nu_3 = a, b, c, d, e$.

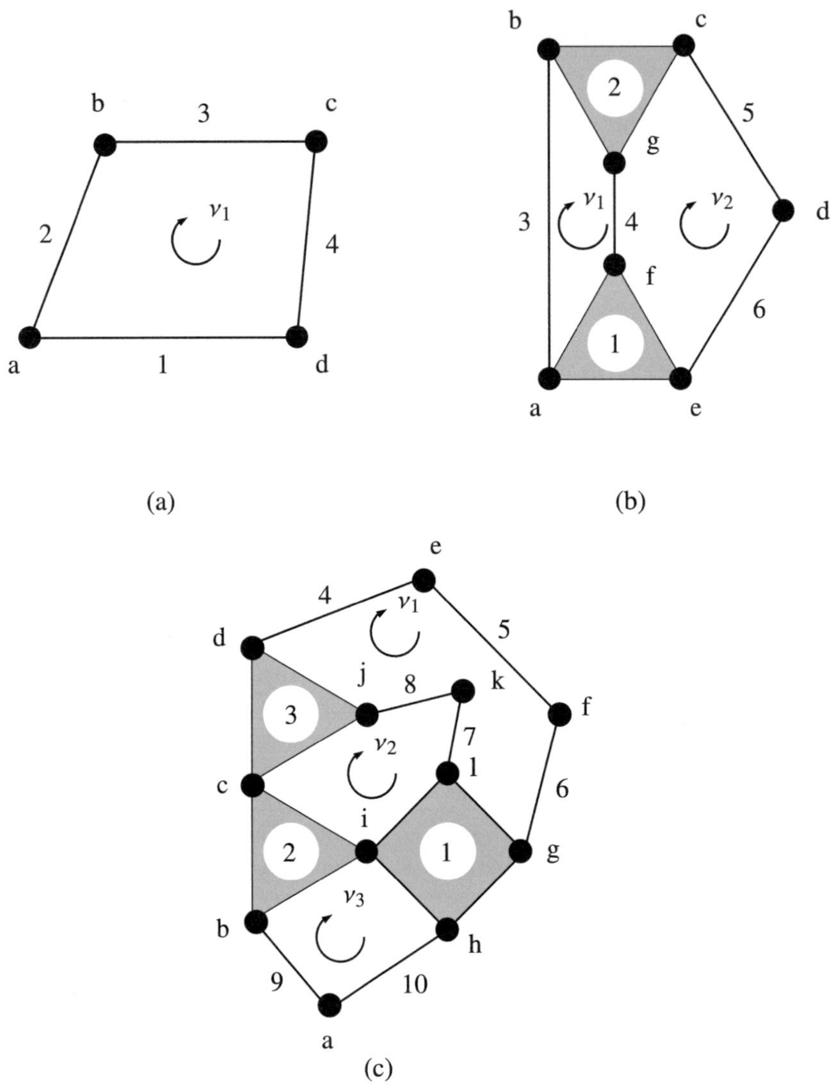

Fig. 1.27 Kinematic chains with varying numbers of independent loops. **a** Kinematic chain with one loop, **b** kinematic chain with two independent loops, **c** kinematic chain with three independent loops

The number of independent loops reflects somehow the degree of complexity of the mechanism, since the higher the value of ν the more components the mechanism will have.

Finally, Fig. 1.27c shows a kinematic chain with three independent loops ($\nu = 3$), these being:

$$\nu_1 = \{d, e, f, g, j, k, l\}, \nu_2 = \{c, i, j, k, l\} \text{ and } \nu_3 = \{a, b, h, i\}$$

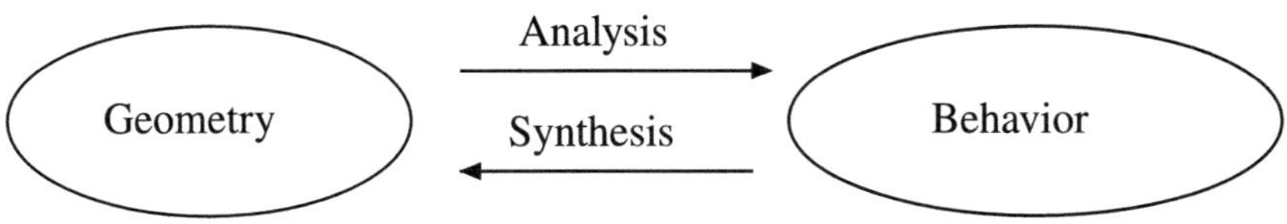

Fig. 1.28 Synthesis and mechanism analysis

1.1.8 Analysis and Synthesis

Analysis is the process of obtaining the behavior of an already designed or existing machine or mechanism. Thus, starting from an existing geometry, the structural characteristics, motions, positions, velocities, accelerations, forces and moments over time are obtained.

Synthesis is the process of obtaining the geometry of a mechanism from the design specifications. Thus, from the design requirements and the desired motions, positions, velocities, accelerations, forces, and moments, the geometry of the mechanism is obtained. The relationship between synthesis and analysis is shown in Fig. 1.28.

Analyzing Theo Jansen's walking mechanism

The Theo Jansen mechanism is again shown in Fig. 1.29. It is desired to determine:

- the representation by graph;
- the number of links;
- the types of each link;
- the number of kinematic pairs;
- the types of each kinematic pair;
- the number of independent loops.

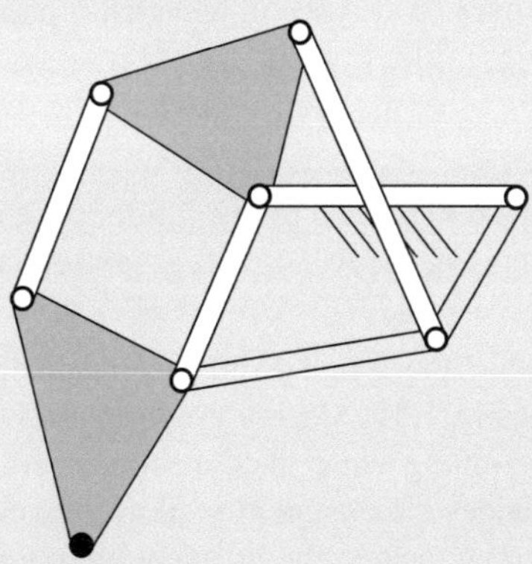

Fig. 1.29 Theo Jansen's walking mechanism

Solution:

Initially, note that kinematic pairs are direct couplings, and by definition couplings are connections between **two and only two** links. Thus, the kinematic pairs that appear highlighted in Fig. 1.30a are two overlapping kinematic pairs. That is, the kinematic pair connecting links 1, 2, and 5 are actually two kinematic pairs, as can be seen in Fig. 1.30b. Analogously, the pair connecting links 3, 5 and 6 and also the pair connecting links 6, 7 and 8 are two overlapping pairs. In Fig. 1.30b, the multiple pairs have been shifted so that it is possible to view all kinematic pairs of the mechanism.

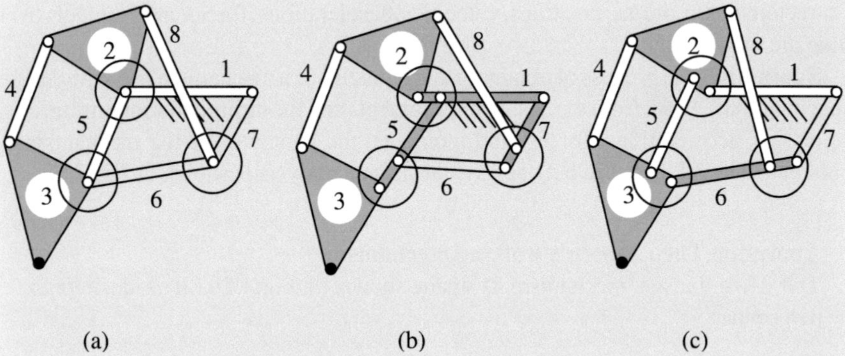

Fig. 1.30 Theo Jansen's mechanism showing all kinematic pairs

There is more than one way to shift these pairs, as can be seen by comparing Figs. 1.30b, c. These two mechanisms have some differences between them, such as the graph, the types of the links, and different motions. However, some of their structural characteristics are the same, such as the number of links, the number of kinematic pairs, the number of loops, the mechanism mobility, the mechanism workspace, and types of kinematic pairs. In addition to the two ways shown in Fig. 1.30b, c, there are other ways to do this kinematic pair change, resulting in other mechanisms, which the reader is encouraged to determine.

The two mechanisms represented in Fig. 1.30b, c are analyzed for didactic purposes. Thus, shown in Fig. 1.31c, f are the graphs of the mechanisms in Fig. 1.30b, c, respectively.

The process to obtain the graph from the mechanism is shown in the sequence of images in Fig. 1.31. Since a graph is just a map of connections between vertices (representing a map of connections between links), the graph is independent of the edges dimension as well as the vertices position. Thus, in order to make the graph clearer, the intersection of edges are avoided. For instance, vertex 1 in Fig. 1.31b was moved to avoid edge crossing, obtaining the Fig. 1.31c. In the graphs obtained, each edge (line between vertices) represents a single kinematic pair (connection between links), the reader is invited to correlate which edge corresponds to which kinematic pair.

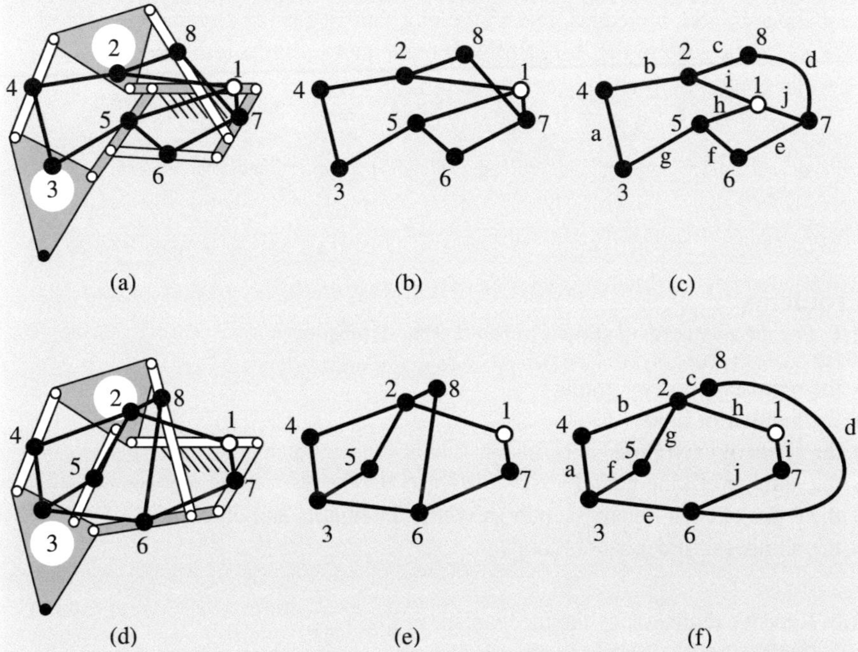

Fig. 1.31 Theo Jansen's mechanism in graph representation showing two different and equivalent ways to represent multiple kinematic pairs

Mechanism of Fig.1.30b

Analyzing the corresponding graph of Fig. 1.31c, one notices that:

- the graph has eight vertices, that is, the mechanism has eight links;
- vertices 3, 4, 6 and 8 have two edges, i.e. links 3, 4, 6 and 8 are binary links. Vertices 1, 2, 5 and 7 have three edges, that is, links 1, 2, 5 and 7 are ternary links;
- the graph has ten edges, that is, the mechanism has ten kinematic pairs;
- as can be seen in Fig. 1.22a, all kinematic pairs are revolute pairs;
- the graph has three independent loops, these can be: $\nu_1 = \{a, b, g, h, i\}$, $\nu_2 = \{c, d, i, j\}$ and $\nu_3 = \{e, f, h, j\}$.

Mechanism of the Fig. 1.30c

Analyzing the graph of Fig. 1.31f, one notices that:

- the graph has eight vertices, that is, the mechanism has eight links;
- vertices 1, 4, 5, 7 and 8 have two edges, i.e. links 1, 4, 5, 7 and 8 are binary links. Vertices 3 and 6 have three edges, i.e. links 3 and 6 are ternary links. Vertex 2 has four edges, that is, link 2 is quaternary;
- the graph has ten edges, that is, the mechanism has ten kinematic pairs;
- as can be seen in Fig. 1.22a, all kinematic pairs are revolute pairs;
- the graph has three independent loops, these can be: $\nu_1 = \{a, b, f, g\}$, $\nu_2 = \{c, d, h, i, j\}$ and $\nu_3 = \{e, f, g, h, i, j\}$.

Problems

1.1 For the mechanism shown in Fig. 1.22b, determine:

- the representation by graph;
- the number of links;
- the types of each link;
- the number of kinematic pairs;
- the types of each kinematic pair (revolute, prismatic, and others);
- the number of independent loops.

1.2 Repeat exercise 1 for the mechanism in Fig. 1.22d.

1.3 Repeat exercise 1 for the mechanism in Fig. 1.23a.

1.4 Repeat exercise 1 for the mechanism in Fig. 1.23b.

1.5 Repeat exercise 1 for the mechanism in Fig. 1.24d.

References

1. Al-Hassan, A.Y., Ahmed, M., Iskandar, A.Z.: Science and Technology in Islam: Technology and Applied Sciences. UNESCO, Paris (2001)
2. Yan, H.S.: Reconstruction Designs of Lost Ancient Chinese Machinery. Springer, Dordrecht (2007)
3. Gupta, S.: Inventions of Archimedes (2014). Available at: http://www.buzzle.com/articles/inventions-of-archimedes.html. Cited 1 Feb 2014
4. Dingley, A.: A Model of a Chinese South-Pointing Chariot (2008). Available at: https://upload.wikimedia.org/wikipedia/commons/c/ce/South-pointing_chariot_(Science_Museum_model).jpg. Cited 7 Feb 2025
5. BBC: O primeiro computador da história? Conheça o objeto mais misterioso da história da tecnologia (2016). Available at: http://www.bbc.com/portuguese/geral-36639213. Cited 5 Aug 2016

6. Ionescu, T.: Terminology for mechanisms and machine science. Mech. Mach. Theory **38**(7–10), 819–825 (2003)
7. Tsai, L.W.: Mechanism Design: Enumeration of Kinematic Structures According to Function. CRC Press, Boca Raton (2000)
8. Tsai, L.W.: Robot Analysis: The Mechanics of Serial and Parallel Manipulators. Wiley, New York (1999)
9. Hunt, K.H.: Kinematic Geometry of Mechanisms. Clarendon Press, Oxford (1978)
10. Hartenberg, R.S., Denavit, J.: Kinematic Synthesis of Linkages. McGraw-Hill, New York (1964)
11. Norton, R.L.: Design of Machinery: An Introduction to the Synthesis and Analysis of Mechanisms and Machines. McGraw-Hill, New York (2008)
12. Tenenbaum, R.A.: Fundamentals of Applied Dynamics. Springer Science & Business Media, New York (2006)
13. Simoni, R.: Síntese estrutural de cadeias cinemáticas e mecanismos. Master's dissertation. Federal University of Santa Catarina, Florianópolis (2008)
14. Fayram, D.: Gakken Strandbeest Rear (2012). Available at: https://www.flickr.com/photos/davefayram/7902830448/. Cited 7 Feb 2025

Chapter 2
Mobility

Abstract This chapter delves into the concept of mobility in mechanical systems, presenting both its theoretical foundations and practical applications. It begins with an explanation of kinematic pair connectivity and constraints, laying the groundwork for understanding how links and joints govern movement. The Gruebler-Kutzbach criterion is introduced as a pivotal equation for calculating degrees of freedom in planar and spatial mechanisms. The chapter discusses how different types of joints—revolute, prismatic, and higher pairs—affect system mobility, and includes tabulated notations for allowed and constrained motions. Special considerations, such as singularities, redundant constraints, and self-alignment, are briefly introduced, highlighting potential design complications. Through a blend of theoretical equations and illustrative figures, the reader gains insight into how mobility impacts mechanical behavior. The chapter ends with practical exercises involving mobility determination in various mechanisms, reinforcing analytical skills essential for mechanism synthesis and design.

Keywords Kinematic pair connectivity · Kinematic pair constraint · Association of kinematic pairs · Mobility equation · Singularity

In this chapter some complementary concepts necessary for the development of the mobility equation is presented. Then the mobility equation is presented for several situations.

2.1 Connectivity and Kinematic Pair Constraint

The **kinematic pair connectivity** is the number of *degrees of freedom* allowed by a given kinematic pair. Kinematic pair connectivity is represented by the symbol f_i, where f comes from *freedom* and the subindex i refers to the kinematic pair i. Analyzing the revolute pair, displayed in Fig. 1.16a, one notices that this kinematic pair allows only one rotation, i.e. , it has kinematic pair connectivity equal to 1. The Table 2.1 presents the notation used for the translational and rotational motions allowed on the axes x, y and z.

D. Martins and E. H. Murai, *Mechanisms–Synthesis and Analysis with Applications in Robotics*, Mechanisms and Machine Science 183,
https://doi.org/10.1007/978-3-031-95187-9_2

Table 2.1 Notations for the allowed and the constrained degrees of freedom of translation (t) and rotation (r)

	rotational			translational		
Degree of freedom	R_x	R_y	R_z	T_x	T_y	T_z
Allowed (f)	f_{rx}	f_{ry}	f_{rz}	f_{tx}	f_{ty}	f_{tz}
Reshetov notation	f''_x	f''_y	f''_z	f'_x	f'_y	f'_z
Davies notation	r	s	t	u	v	w
Constrained (c)	c_{rx}	c_{ry}	c_{rz}	c_{tx}	c_{ty}	c_{tz}
Reshetov notation	M_x	M_y	M_z	Q_x	Q_y	Q_x
Davies notation	R	S	T	U	V	W

The **kinematic pair constraint** is the number of constraints imposed by a given kinematic pair. The kinematic pair constraint is represented by the symbol c_i, where c comes from *constraint* and the subindex i refers to the kinematic pair i. Analyzing the revolute pair, displayed in Fig. 1.16a, one notices that this kinematic pair constrains three translations and two rotations, i.e. it has kinematic pair constraint equal to 5. Table 2.1 presents the notation used for the translation and rotation constraints on the x, y and z axes. Besides the notation presented in Table 2.1, other notations are found in the literature, such as the notation of [1] for actions, with forces represented by Q_x, Q_y, Q_z and moments represented by M_x, M_y, M_z. Another notation proposed by [2] has the moments at x, y and z represented respectively by R, S, T, and the forces at x, y and z represented respectively by U, V, W. Whereas for motions, rotations at x, y, and z are represented respectively by r, s, t, and translations at x, y, and z are represented respectively by u, v, w.

A free body in space has six possible independent motions[1], the three translations and the three rotations. These six possible degrees of freedom can be constrained or not. In a kinematic pair, by restricting one degree of freedom, the same degree of freedom starts to count as the kinematic pair contraint. In that same pair, by not restricting a degree of freedom, that unrestricted degree of freedom counts as a kinematic pair connectivity. Thus, the two concepts presented above, kinematic pair connectivity and kinematic pair constraint, are complementary concepts. So, for each kinematic pair

$$f_i + c_i = \lambda \tag{2.1}$$

The symbiotic relationship between connectivity and constraint is humorously captured in the phrase: "What comes loose is what does not bind, and what binds is what does not come loose." This tongue-in-cheek expression encapsulates the intricate balance between these two concepts. Table 2.2 illustrates the connectivity and constraint of kinematic pairs for lower pairs. Table 2.2 also shows the amount

[1] Strictly mathematically, a motion in space can be represented as a linear combination of the three translations f_{tx}, f_{ty} and f_{tz} and the three rotations f_{rx}, f_{ry} and f_{rz}. Thus, these six motions, being independent of each other, can be used to assemble the basis of the vector space of motions in space

Table 2.2 Connectivity and kinematic pair constraint. Here, NP is non-planar, so the kinematic pair is not viable for $\lambda = 3$

Kinematic pair	Motion	c_i ($\lambda = 3$)	Class	c_i ($\lambda = 6$)	f_i
Revolute	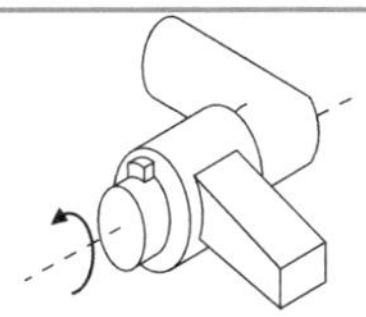	2	V	5	1
Prismatic	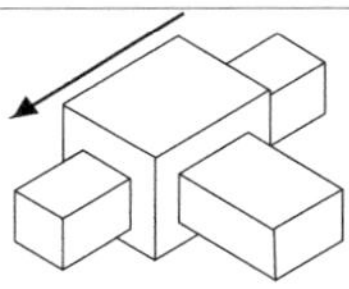	2	V	5	1
Helical	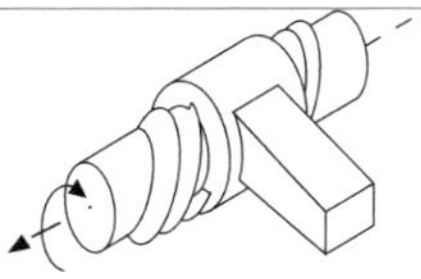	NP	V	5	1
Cylindrical	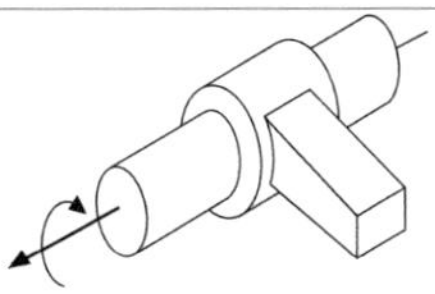	NP	IV	4	2
Planar	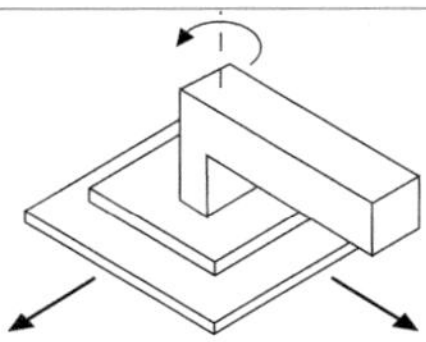	0	III	3	3
Spherical	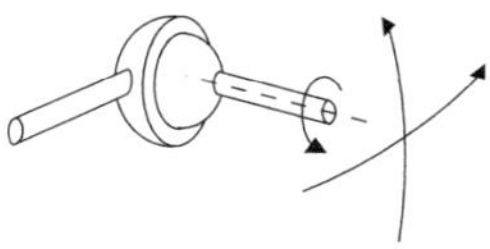	NP	III	3	3

of constraints that each kinematic pair presents considering the spatial workspace ($\lambda = 6$) and the planar workspace ($\lambda = 3$), also the class of kinematic pairs according to [1] is presented.

In general, lower pairs are characterized by the type and number of freedoms (f_i) while higher pairs are normally characterized by the type and number of constraints (c_i)

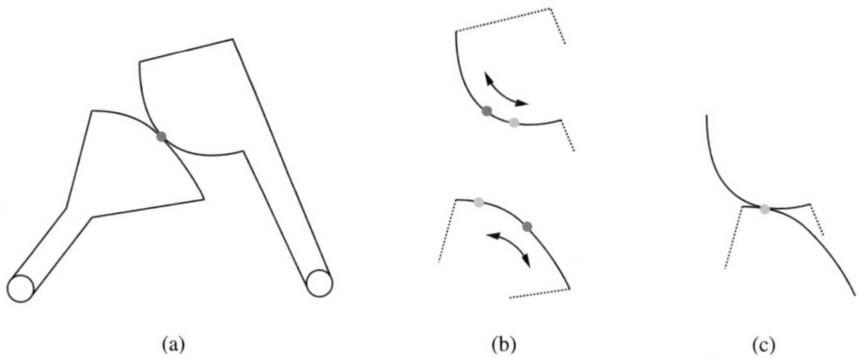

Fig. 2.1 Connectivity and kinematic pair constraint on a higher pair

In higher kinematic pairs, the analysis of kinematic pair connectivity and kinematic pair constraint is not as intuitive. Consider the linear contact kinematic pair, shown in Fig. 2.1a. This pair connects two links through a contact line that passes through the highlighted point. Considering only this higher kinematic pair in isolation, it is noted that the contact point must be on the surface of both links. Thus, considering only one of the links, the contact point can move over the surface of the link reaching various positions (see Fig. 2.1b).

Considering the other link, the contact point can also move over the surface of the link, reaching various positions (see Fig. 2.1b). Having chosen the two new points, you can join the kinematic pair in such a way that the contact point coincides with the two chosen points (see Fig. 2.1c). You have two degrees of freedom in this kinematic pair, as you can choose the position of one point on one surface and the other point on the other surface. So this pair has connectivity two.

Another way to determine the connectivity of this kinematic pair is to consider the possible motions that one link can make relative to the other. For this analysis, one must remember that the kinematic pair must exist, i.e. , the two *elements of kinematic pair* must be in contact. By fixing one of the links in Fig. 2.1a and moving the other, it is possible to make the contact point move over the surface of the links, as can be seen in Fig. 2.2a. Also, keeping the contact point in a constant position, one of the links can be rotated around the contact point, as can be seen in Fig. 2.2b.

Thus, the kinematic pair in Fig. 2.1a allows two motions, a translation and a rotation, where the translation refers to the translation of the contact point over the surface of the link and the rotation refers to the rotation of the link around the contact point.

At a time instant the higher kinematic pair in Fig. 2.3a can be replaced by two lower pairs, as shown in Fig. 2.3b. As discussed in the section 1.1.4, the replacement of a kinematic pair with two degrees of freedom ($f_i = 2$) by two kinematic pairs with one degree of freedom ($f_i = 1$) is called *expansion of kinematic pair*. In Fig. 2.3a, the arrow indicates the translation direction at the time instant.

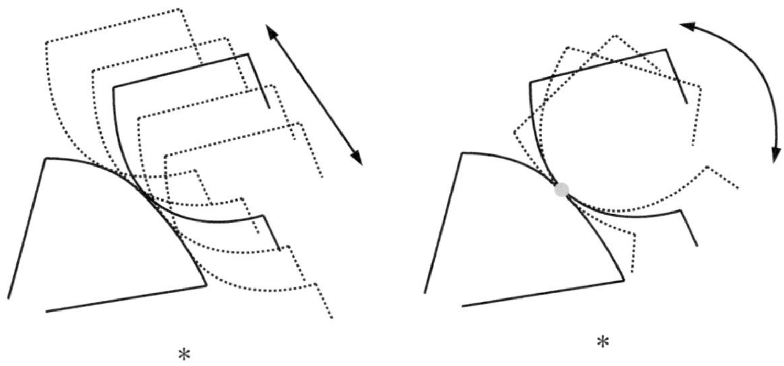

(a) (b)

Fig. 2.2 Connectivity of a higher kinematic pair. **a** translation degree of freedom. **b** rotation degree of freedom

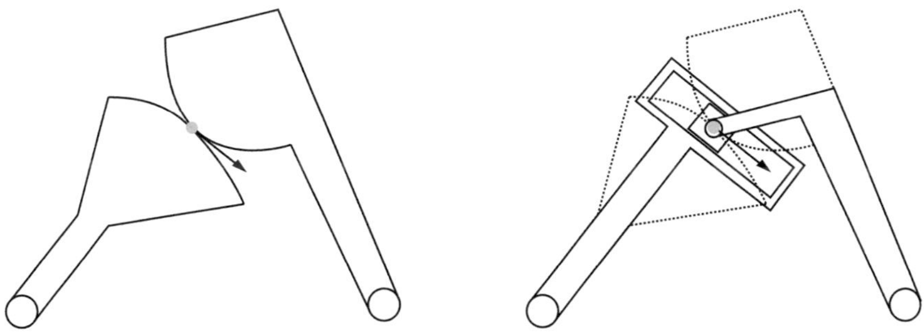

(a) (b)

Fig. 2.3 Expansion in lower kinematic pairs

When doing the expansion, the prismatic kinematic pair must have the same direction as the arrow, so in both Figs. 2.3a and 2.3b, the translation occurs in the same direction. Similarly, in the case of Fig. 2.3a the rotation occurs around the point at the base of the arrow. Thus, for Fig. 2.3b a revolute joint must be inserted about this point. Therefore, both Figs. 2.3a and 2.3b present the same two motions, showing that the higher kinematic pair in question has $f_i = 2$. Finally, it is noted that this kinematic pair expansion is valid only at a given time instant. For another time instant, the prismatic pair must be reoriented and the revolute pair must be repositioned.

More content regarding connectivity and degree of freedom in kinematic pairs can be seen at [1, 3–5].

2.2 Series and Parallel Association of Kinematic Pairs

Kinematic chains are generated through the connection between links, and such connection is done by kinematic pairs. Thus, a kinematic chain can be considered as an association of kinematic pairs in series or in parallel.

2.2.1 *Association of Kinematic Pairs in Series*

Kinematic pairs can be associated in series, as seen in Fig. 2.4.

In Fig. 2.4a there are three separate kinematic pairs that will be associated, being one revolute pair and two prismatic pairs. Associating these pairs in series results in the open kinematic chain in Fig. 2.4b. Note that the point *P* can translate in x and y and rotate in z with respect to the ground, i.e. , it has f_{tx}, f_{ty} and f_{rz}. These degrees of freedom that *P* has with respect to the ground are the combined degrees of freedom of each kinematic pair. Thus, **in the series association the degrees of freedom are the union of the degrees of freedom of each kinematic pair** ($f = \cup f_i$).

Considering the constraints, the revolute pair has c_{tx} and c_{ty}, the translation pair along x has the constraints c_{rz} and c_{ty} and the translation pair along y has the constraints c_{rz} and c_{tx}. But notice that when serially associating, the point *P* has all the degrees of freedom of the plane, so it does not have any constraints of the plane. Thus, **in serial association the constraints are the intersection of the constraints of each kinematic pair**, ($c = \cap c_i$). All kinematic pairs have the constraints c_{tz}, c_{rx} and c_{ry}, so the resulting mechanism will be planar.

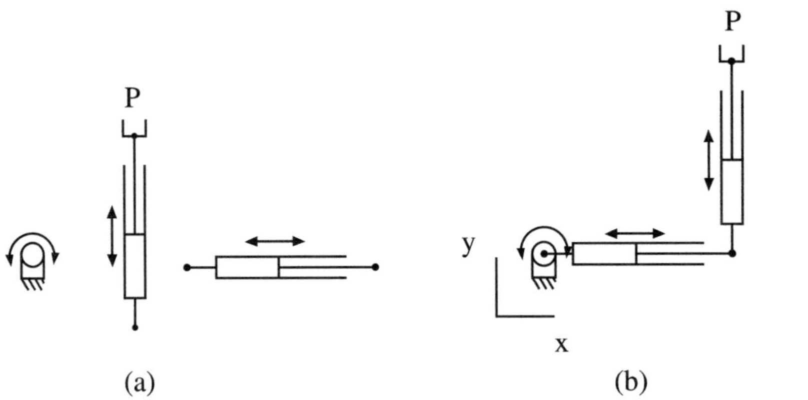

Fig. 2.4 Serial association of kinematic pairs

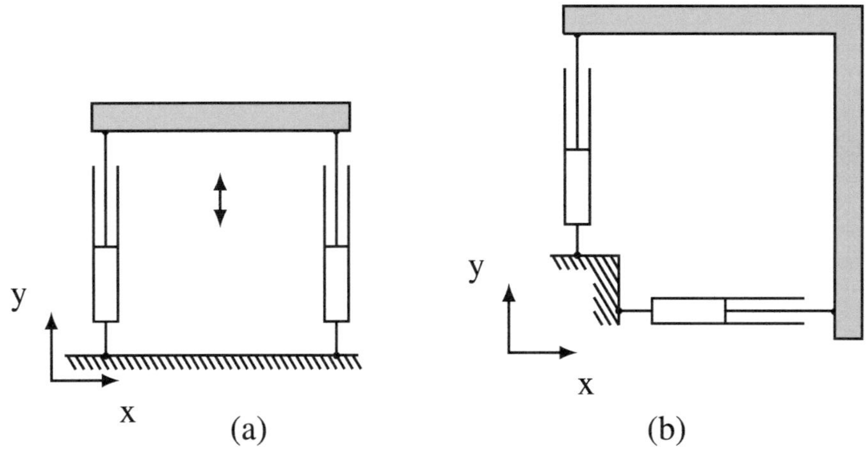

Fig. 2.5 Parallel association of kinematic pairs

2.2.2 *Association of Kinematic Pairs in Parallel*

Kinematic pairs can also be associated in parallel, as shown in Fig. 2.5.

In Fig. 2.5a, the two prismatic pairs are positioned so that both allow displacement f_{ty}. Note that the platform allows only f_{ty} displacement. In Fig. 2.5b, the prismatic pairs have been positioned so that one of them allows displacement f_{tx} and the other f_{ty}. Note that the platform has no degree of freedom, because when trying to translate it in x, the prismatic pair in y prevents such motion. Similarly, when trying to translate the platform at y, the prismatic pair at x prevents the motion. Thus, **in the parallel association the degrees of freedom are the intersection of the degrees of freedom of each kinematic pair**. In a complementary way, one notes that **in parallel association the constraints are the union of the constraints of each kinematic pair**, ($f = \cap f_i$) and ($c = \cup c_i$).

In section 1.1.4, some types of kinematic pairs were shown, among them the point-contact kinematic pair, in Fig. 1.17c. This kinematic pair, shown again in Fig. 2.6a, has five degrees of freedom (f_{tx}, f_{ty}, f_{rx}, f_{ry} and f_{rz}) and one constraint (c_{tz}). By conveniently associating two point pairs in parallel, two constraints are obtained, each of which comes from a point contact. Such an association is shown in Fig. 2.6b, whose association generates constraints c_{ty} and c_{tz}. Similarly, one can associate three point pairs in parallel, obtaining the constraints c_{tx}, c_{ty} and c_{tz}. Such an association is shown in Fig. 2.6c, where the result is equivalent to a spherical kinematic pair, since it has only f_{rx}, f_{ry} and f_{rz}.

The three kinematic pairs shown in Fig. 2.6 can be associated again in parallel in order to obtain all the constraints, c_{tx}, c_{ty}, c_{tz}, c_{rx}, c_{ry} and c_{rz}. This association is displayed in Fig. 2.7. Figure 2.7a shows the disconnected kinematic pair elements, and Fig. 2.7b shows the connected elements forming a Kelvin coupling [6].

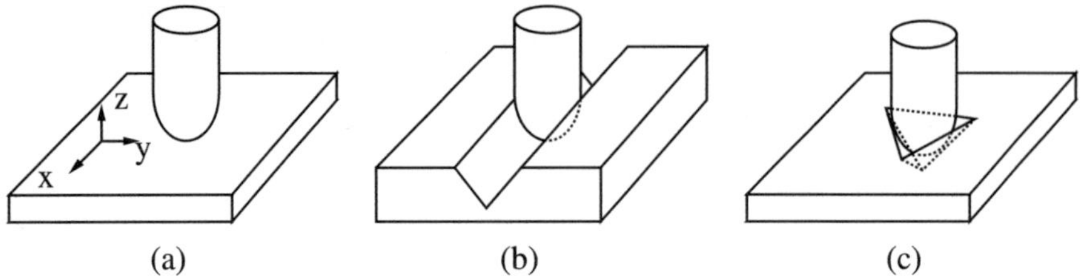

Fig. 2.6 Parallel association of point kinematic pairs. **a** point contact. **b** two point contacts. **c** three point contacts

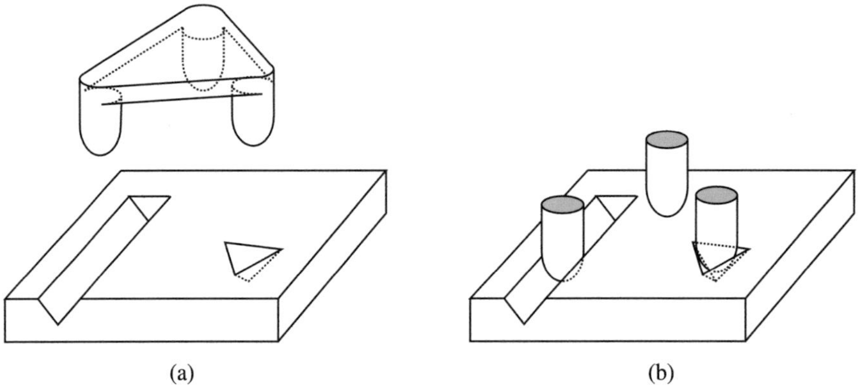

Fig. 2.7 Parallel association of point kinematic pairs excluding all motions. **a** separated Kelvin coupling. **b** assembled Kelvin coupling

The kinematic chain in Fig. 2.7 constrain any motion between the two links, achieving precision in the fit between the links and also speed in assembly and disassembly. Such a kinematic chain is used in tribrachs, which are devices for mounting theodolites and global positioning systems on tripods.

2.3 Mobility Equation

This section develops the arguably the most important sets of equations of the theory of mechanisms.

Consider a free body in planar workspace, as indicated in Fig. 2.8.

In planar workspace, a body has the motions f_{tx}, f_{ty} and f_{rz}. Any other motions that are linear combinations of the previous three motions are also allowed. Note that a body in this space will always possess the constraints c_{tz}, c_{rx} and c_{ry}, since these are the three constraints that generate the planar space. Thus, in a mechanical system whose workspace is planar, these three constraints are usually not accounted for, so that a free body in this planar space has constraints $c_i = 0$.

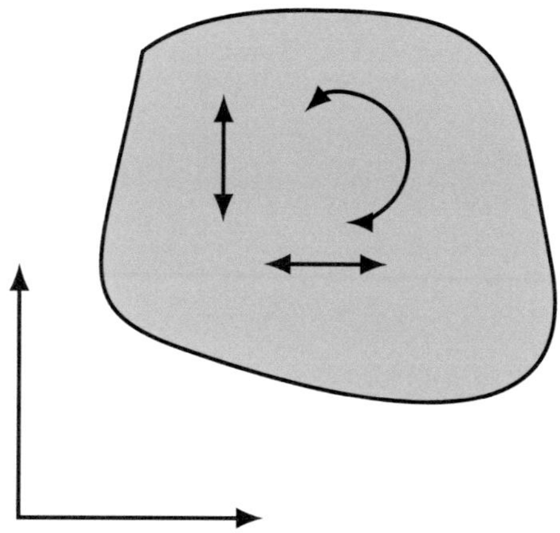

Fig. 2.8 body in planar workspace

A free body in planar space ($\lambda = 3$, as seen in section 1.1.7) has three degrees of freedom, $f_i = 3$. Therefore, by Eq. 2.1, we have that $c_i = 0$. Thus, such a body has no constraints c_{tx}, c_{ty} and c_{rz} (constraints belonging to the workspace), but it has the constraints c_{tz}, c_{rx} and c_{ry} (constraints not belonging to the workspace), which are usually not accounted for because they are always present.

A mechanical system consisting of two free bodies in planar space (see Fig. 2.9a) has six degrees of freedom, these being f_{tx}, f_{ty} and f_{rz} for both bodies. Connecting the bodies with a contact point pair (as in Fig. 2.1) results in the mechanical system of Fig. 2.9b. The contact point pair has two degrees of freedom in the planar workspace (f_{tx} and f_{rz}), allowing two degrees of freedom between the two bodies. This contact point pair in the planar workspace has connectivity 2 and constraint 1 ($f_i = 2$ and $c_i = 1$). The resulting mechanical system has five degrees of freedom, three for positioning the *1* body (f_{tx}, f_{ty} and f_{rz}) and two for positioning the *2* body relative to the *1* body (f_{tx} and f_{rz}).

The mechanical system in Fig. 2.9 initially had 2λ degrees of freedom. When the contact point pair is inserted, it introduces c_i constraint, and the number of degrees of freedom drops to $2\lambda - c_i$. Also, if one link is fixed, all possible motions for that link within the workspace are removed, i.e. , λ degrees of freedom are removed.

The **mobility** (M) of a kinematic chain or mechanism is the number of independent parameters needed to define the configuration of the kinematic chain or mechanism. In other words, mobility is the degree of freedom from relative (internal) motion of the kinematic chain or mechanism. Thus, for a free body in the plane, the mobility is

$$M = \lambda.$$

For a mechanical system with two free bodies in the plane (Fig. 2.9a)

$$M = 2\lambda.$$

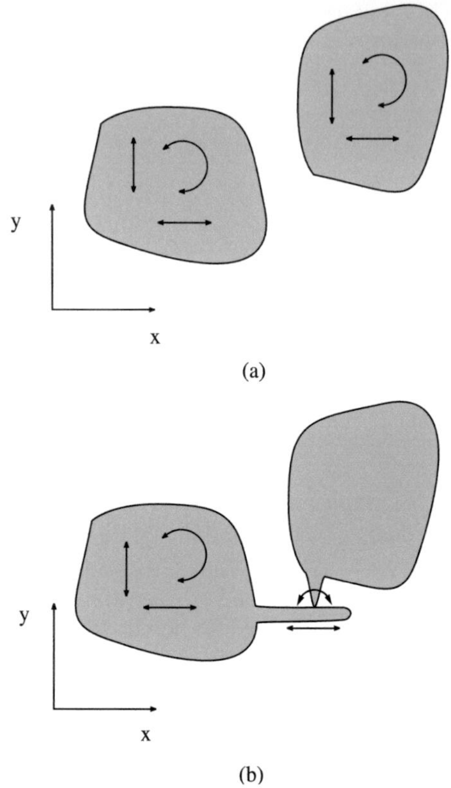

Fig. 2.9 Two bodies in the plane

For n free bodies in the plane

$$M = n\lambda.$$

In mechanism science, all absolute referential is a moving referential; therefore, we fix one of the bodies, choosing one body as the absolute referential

$$M = (n - 1)\lambda.$$

If we insert p kinematic pairs connecting the bodies, adding constraints, we must subtract from the mobility the sum of all constraints that have been inserted

$$M = (n - 1)\lambda - \sum_{i=1}^{p} c_i. \tag{2.2}$$

This is the first and more generic mobility equation which is particularly useful for cases where all or almost all of the kinematic pairs are higher pairs which are mainly characterized by their constraints c_i.

However for most cases there is a need for simpler mobility equations.

So, substituting Eq. 2.1

$$M = (n-1)\lambda - \sum_{i=1}^{p} (\lambda - f_i)\,.$$

and manipulating the expression

$$M = (n-1)\lambda - \sum_{i=1}^{p} \lambda + \sum_{i=1}^{p} f_i$$

$$M = (n-1)\lambda - p\lambda + \sum_{i=1}^{p} f_i$$

$$M = (n-1-p)\lambda + \sum_{i=1}^{p} f_i\,. \tag{2.3}$$

Recall that a kinematic pair of connectivity greater than one can be expanded into multiple kinematic pairs with connectivity one (see *kinematic pair expansion*, in section 1.1.4). The number of kinematic pairs in a chain with all kinematic pairs expanded is represented by the symbol j. Thus, the mobility equation can be rewritten for expanded kinematic pairs

$$M = (n-1-j)\lambda + \sum_{i=1}^{j} f_i\,.$$

Since the kinematic pairs are expanded, every kinematic pair has $f_i = 1$. Thus, the summation of all f_i results in the number of kinematic pairs j.

$$M = (n-1-j)\lambda + j\,. \tag{2.4}$$

Equation 2.3 should be used for **mechanisms with unexpanded kinematic pairs**. Equation 2.4 should be used for **mechanisms with expanded kinematic pairs**.

More information regarding the mobility equation can be seen at [4, 7–12].

An additional frequently used equation is Euler's equation[2], developed for graphs and which can be used to relate the number of independent loops, links, and kinematic pairs.

$$\nu - 1 = j - n\,. \tag{2.5}$$

[2] Leonhard Euler (1707-1783), Swiss mathematician and physicist known for his work in mathematical analysis, graphs, mechanics, fluid dynamics, optics, astronomy, and music theory

Table 2.3 Mobility equations

	Unexpanded pairs	Expanded pairs
Regular	$M = (n-1-p)\lambda + \sum_{i=1}^{p} f_i$	$M = (n-1-j)\lambda + j$
Including Euler	$M = -\nu\lambda + \sum_{i=1}^{p} f_i$	$M = -\nu\lambda + j$

Note that when a kinematic pair is expanded, more links and pairs are added in series. Regarding the graph representation, the expansion adds vertices and edges in a serial manner, increasing n and j equally. However, the number of independent loops is not affected, and neither is the right-hand side of Eq. 2.5. Therefore, Euler's equation can be used for kinematic chains with expanded and unexpanded pairs.

Substituting Euler's equation into Eqs. 2.3 and 2.4, it results, respectively,

$$M = -\nu\lambda + \sum_{i=1}^{p} f_i \tag{2.6}$$

and

$$M = -\nu\lambda + j. \tag{2.7}$$

To summarize, the mobility equations are presented in Table 2.3.

Mechanisms mobility calculation

Determine the mobility of the planar mechanisms in Figs. 2.10a, 2.10b and 2.10c.

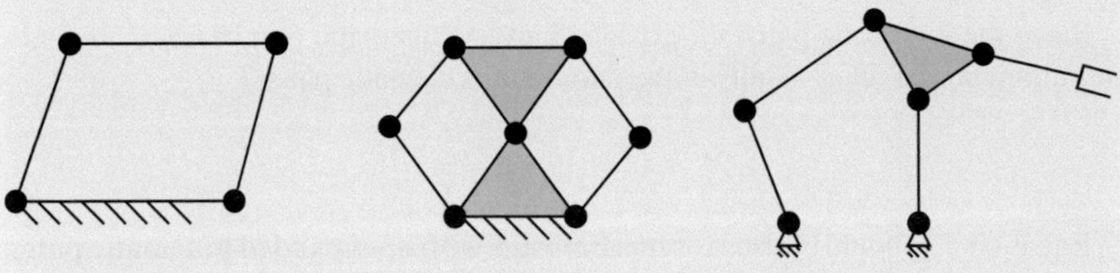

(a) (b) (c)

Fig. 2.10 Planar mechanisms

Soution:

Mechanism of the Fig. 2.10a

The mechanism shown in Fig. 2.10a features four links and four kinematic pairs with connectivity one. Also, the mechanism is planar, so $\lambda = 3$. Substituting into Eq. 2.4, one has

$$M = (n - 1 - j)\lambda + j = (4 - 1 - 4)3 + 4 = 1.$$

Thus, the mechanism mobility is 1.

Mechanism from Fig. 2.10**b**

The mechanism shown in Fig. 2.10b has six links and seven kinematic pairs with connectivity one. Since the mechanism is planar, one has $\lambda = 3$. Substituting into Eq. 2.4, one has

$$M = (n - 1 - j)\lambda + j = (6 - 1 - 7)3 + 7 = 1.$$

Thus, the mechanism mobility is 1.

Mechanism from Fig. 2.10**c**

The mechanism shown in Fig. 2.10c features six links and six kinematic pairs with connectivity one. Since the mechanism is planar, $\lambda = 3$. Substituting in Eq. 2.4, one has

$$M = (n - 1 - j)\lambda + j = (6 - 1 - 6)3 + 6 = 3.$$

Thus, the mechanism mobility is 3, corresponding to two degrees of freedom that must be set to determine the closed subchain and one degree of freedom to set the claw position.

Planar mechanism mobility calculation

Determine the mobility of the planar mechanism in Fig. 2.11. The mechanism is composed of three revolute pairs and one planar kinematic pair E.

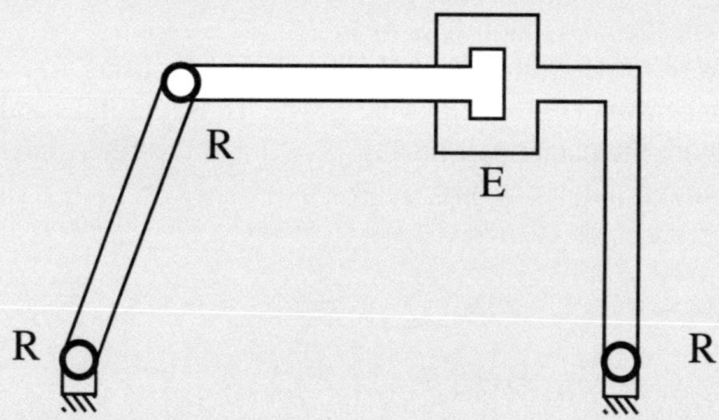

Fig. 2.11 Mechanism with three revolute pairs (R) and one planar pair (E)

Solution:

The analysis should be done with caution. In Fig. 2.11, the two revolute pairs that are fixed present one link, link 1 indicated in Fig. 2.12a. To use Eq. 2.4, one must expand kinematic pairs that have pair connectivity greater than 1. The planar pair has connectivity three ($f_i = 3$), so one must expand it into three pairs of connectivity one while retaining all the degrees of freedom of the pair. Since the planar pair allows two translations and one rotation, when expanding one must use two prismatic pairs and one revolute pair, as shown in Fig. 2.12b. The circle in Fig. 2.12b shows the three kinematic pairs resulting from expanding the planar pair.

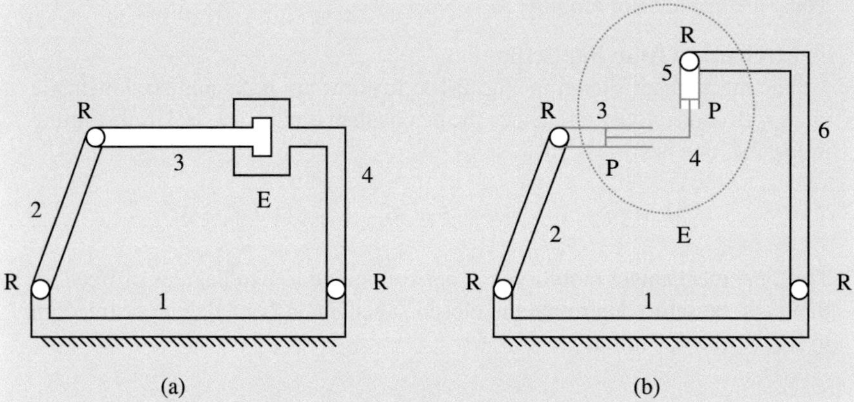

Fig. 2.12 Mechanism with planar pair. **a** contracted planar pair. **b** expanded planar pair

Analyzing Fig. 2.12b, one can see that the mechanism has six links and six kinematic pairs. Substituting into Eq. 2.4, one has

$$M = (n - 1 - j)\lambda + j = (6 - 1 - 6)3 + 6 = 3.$$

Thus, the mechanism mobility is $M = 3$.

Another way to determine the mobility would be using Eq. 2.3. In this case, the mechanism in Fig. 2.12a is used, which has four links and four kinematic pairs, with the planar pair having $f_i = 3$ and the others having $f_i = 1$. Substituting into Eq. 2.3, one has

$$\begin{aligned} M &= (n - 1 - p)\lambda + \sum_{i=1}^{p} f_i = (4 - 1 - 4)3 + \sum_{i=1}^{4} f_i \\ &= (4 - 1 - 4)3 + (1 + 1 + 1 + 3) = 3. \end{aligned}$$

As expected, the same result $M = 3$ is obtained.

Determining the mobility in a mechanism with higher kinematic pairs

Determine the mobility of the planar mechanism in Fig. 2.13. The mechanism is composed of five contact point kinematic pairs, as shown in Fig. 2.1a.

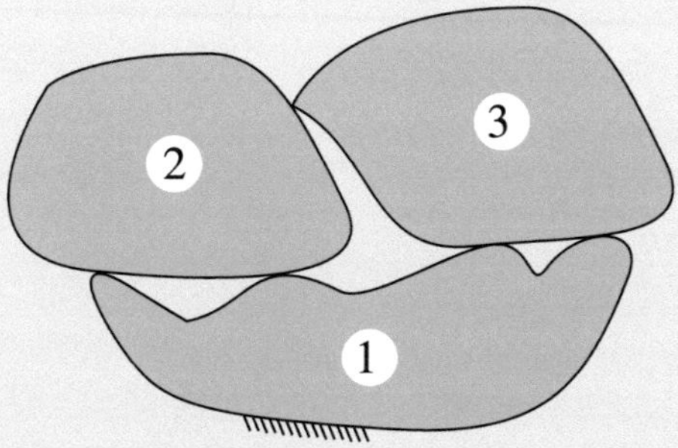

Fig. 2.13 Planar mechanism with five contact point pairs

Solution:

As shown in the section 2.1, the point pair in the planar workspace has kinematic pair connectivity two, because such a pair has two degrees of freedom. One can expand the kinematic pair by replacing the point pair with two pairs with one degree of freedom each. The kinematic pair expansion is shown in Fig. 2.14. In Fig. 2.14, the kinematic pairs shown only represent one degree of freedom, not necessarily being a revolute pair. Thus, the pairs shown in Fig. 2.14 only serve to highlight the individualization of each degree of freedom.

By expanding all the contact point kinematic pairs, one has a mechanism with eight links and ten kinematic pairs with one degree of freedom each (mechanism shown in Fig. 2.14c). Thus, from Eq. 2.4 one has

$$M = (n - 1 - j)\lambda + j = (8 - 1 - 10)3 + 10 = 1.$$

Another way to determine the mobility would be using Eq. 2.3. In this case, the mechanism in Fig. 2.13 is used, whose mechanism has three links and five kinematic pairs, each pair with $f_i = 2$. Substituting into Eq. 2.3, one has

$$\begin{aligned} M &= (n - 1 - p)\lambda + \sum_{i=1}^{p} f_i = (3 - 1 - 5)3 + \sum_{i=1}^{5} f_i \\ &= (3 - 1 - 5)3 + (2 + 2 + 2 + 2 + 2) = 1. \end{aligned}$$

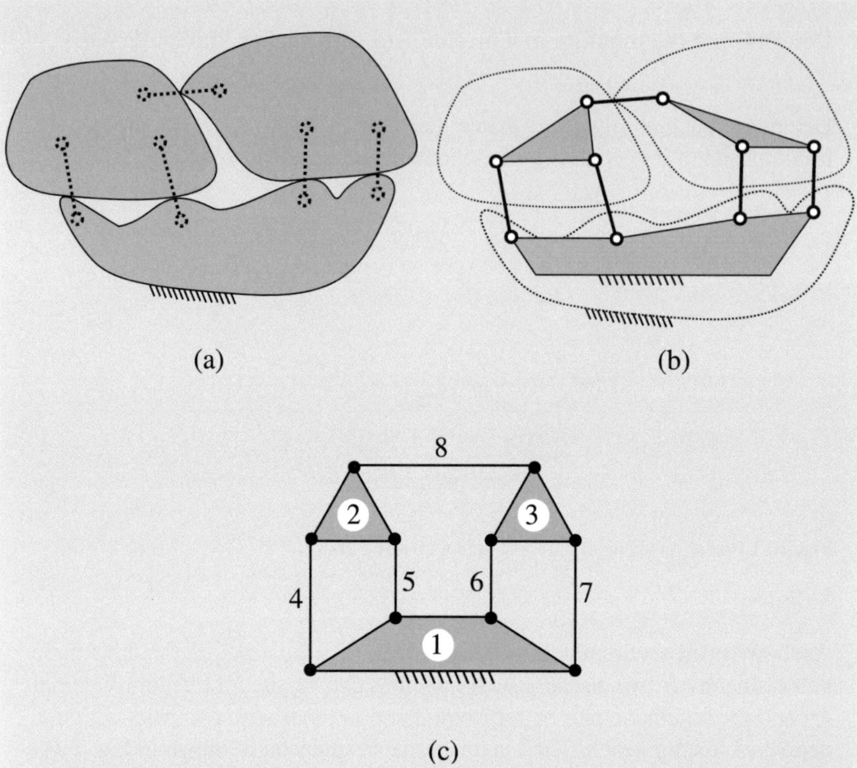

Fig. 2.14 Planar mechanism with five contact point pairs. **a** original mechanism. **b** mechanism with expanded kinematic pairs. **c** structural representation of the mechanism with expanded kinematic pairs

2.4 Singularity

By actuating the mechanism, it can enter some positions of **singularity**. In such positions the kinematics of the mechanism is changed, and it may temporarily lose or gain some degrees of freedom. Upon leaving the position of singularity, the mechanism returns to its usual mobility.

This is because the mobility equations, deduced in this chapter, consider only the quantities of elements and degrees of freedom. Thus, the mobility equation disregards the way the links are connected: the types of kinematic pairs and the motion allowed by each degree of freedom, i.e. , whether it is a rotation around x or y.

For example, the space robot in Fig. 2.15a has mobility 3, and can rotate in θ_1, θ_2 and θ_3. In the situation depicted in Fig. 2.15a, these three rotations allow the robotic

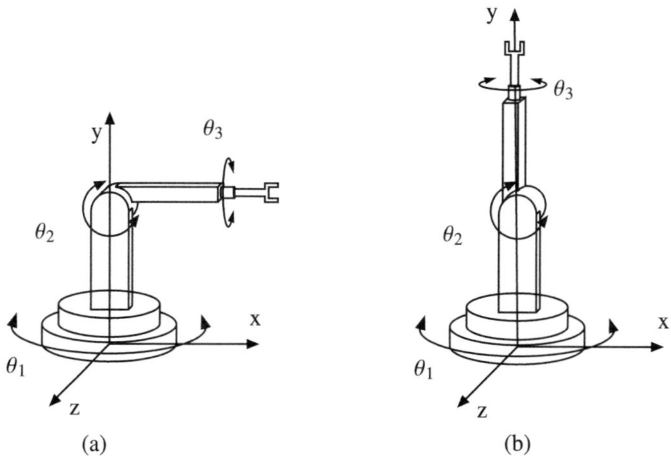

Fig. 2.15 Singularity reducing the degree of freedom

gripper to have three degrees of freedom with respect to the base. In general, these three degrees of freedom predominate.

However, in Fig. 2.15b, even though the robot has mobility 3, the gripper only has two degrees of freedom with respect to the base, because θ_1 and θ_3 represent rotation about the same axis, y. Thus, for this position, the gripper can only make two distinct movements, having lost one degree of freedom.

Another example can be seen in Fig. 2.16a. In this figure, the four-bar mechanism exhibits mobility one, which holds for any given position. However, the degree of freedom is increased when the alignment of links shown in Fig. 2.16b occurs.

Note that in this case, when rotating the left link clockwise, there are two possibilities for the right link, it can rotate clockwise or counterclockwise. So to define which direction of rotation, one needs, at this instant of time, to drive the right link in the desired direction, increasing the mobility of the mechanism.

Four-bar mechanisms are designed to avoid this alignment, because it is not known which direction the right-hand link will go. This design requirement will be seen later in Chaps. 7 and 8.

The origin of the term singularity refers to the null value of the determinant of the Jacobian matrix (a matrix that will be seen in Chap. 5). The Jacobian matrix describes algebraically the geometry of the mechanism. Thus, particular geometric positions can nullify the determinant of the Jacobian matrix. Among such positions are the alignment and parallelism of axes. Thus, the designer should be aware of geometric positions that are particular cases.

Some authors consider the correct term for singularity (an analysis-based term) is *stationary configuration* (a kinematic-based term).

There are several types of singularity, however it is not the scope of this book to analyze the origin and effect of each type of singularity. What the reader should keep in mind is that in certain positions, the mobility equation may not hold.

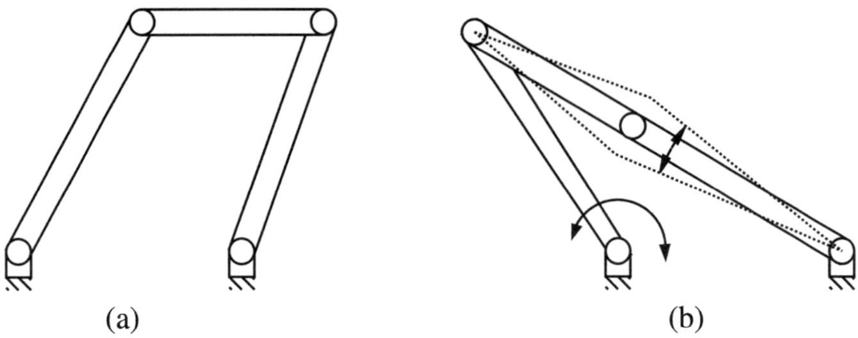

Fig. 2.16 Singularity increasing the degree of freedom

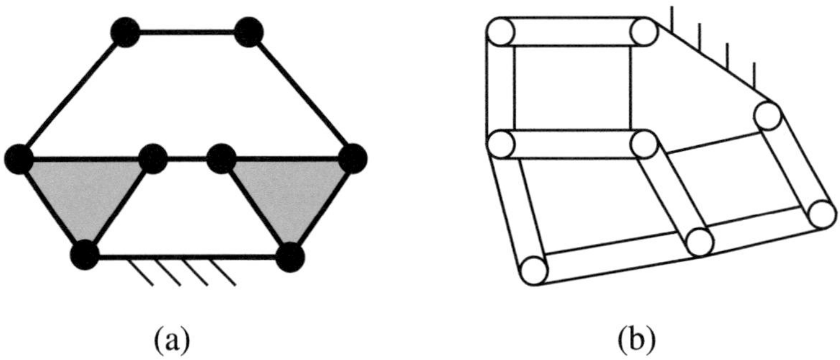

Fig. 2.17 Planar mechanisms with two and three independent loops

Another quite important topic is the presence of redundant constraints and self-alignment but these topics go a little beyond the scope of this book.

Problems

2.1 Determine the mobility of the mechanism shown in Fig. 1.22b.

2.2 Determine the mobility of the mechanism shown in Fig. 1.22d.

2.3 Determine the mobility of the mechanism shown in Fig. 1.23a.

2.4 Determine the mobility of the mechanism shown in Fig. 1.23b.

2.5 Determine the mobility of the mechanism shown in Fig. 1.24d.

2.6 Determine the mobility of the mechanism shown in Fig. 2.17a.

2.7 Determine the mobility of the mechanism shown in Fig. 2.17b.

References

1. Reshetov, L.: Self-aligning Mechanisms. Mir Publishers, Moscou (1982)
2. Davies, T.: Freedom and constraint in coupling networks. J. Mech. Eng. Sci.: Part C, **220**(7), 989–1010 (2006)
3. Hartenberg, R.S., Denavit, J.: Kinematic Synthesis of Linkages. McGraw-Hill, New York (1964)
4. Hunt, K.H.: Kinematic Geometry of Mechanisms. Clarendon Press, Oxford (1978)
5. Martins, D., Carboni, A.P.: Variety and connectivity in kinematic chains. Mech. Mach. Theory **43**(10), 1236–1252 (2008)
6. Ding, J., Liu, X., Wang, C.: Repeatability analysis of an overconstrained kinematic coupling using a parallel-mechanism-equivalent model. IEEE Access **9**, 82455–82461 (2021)
7. Norton, R.L.: Design of Machinery: An Introduction to the Synthesis and Analysis of Mechanisms and Machines. McGraw-Hill, New York (2008)
8. Suh, C., Radcliffe, C.: Kinematics and Mechanisms Design. R.E. Krieger Pub. Co., Malabar (1978)
9. Gogu, G.: Mobility of mechanisms: a critical review. Mech. Mach. Theory **40**(9), 1068–1097 (2005)
10. Tsai, L.W.: Mechanism Design: Enumeration of Kinematic Structures According to Function. CRC Press, Boca Raton (2000)
11. Mabie, H.H., Reinholtz, C.F.: Mechanisms and Dynamics of Machinery. Wiley, Chichester (1987)
12. Phillips, J.: Freedom in Machinery: Introducing Screw Theory. Cambridge University Press, Cambridge (1984)

Part II
Mechanism Analysis

Chapter 3
Four-Bar Linkages

Abstract This chapter focuses exclusively on four-bar linkages, one of the most fundamental and widely used planar mechanisms. It introduces the different classifications of four-bar linkages—crank-rocker, double-crank, double-rocker—based on link lengths and motion behavior. The Grashof condition is presented to determine the possible type of motion within a given configuration. Various configurations and their input-output relationships are analyzed through geometric and trigonometric principles. Readers are guided through the loop-closure equation, a key tool for analyzing linkage positions and deriving transmission quality equation. The chapter highlights the versatility of the four-bar mechanism, noting its applications in tools, robotics, and manufacturing systems. Clear illustrations support the theoretical discussions, helping visualize link positions and angular constraints. Practical design implications, such as selecting appropriate link lengths to achieve continuous or oscillatory motion, are emphasized. Overall, the chapter provides a robust introduction to a central element of mechanism design.

Keywords Four-bar linkage · Grashof criterion · Transmission quality · Transmission angle

In this chapter, the four-bar mechanism is studied. Initially some complementary concepts are presented, then Grashof's criterion and finally transmission quality and angle concepts are discussed. For more information regarding four-bar linkages, see [1–3].

3.1 Preliminary Concepts on Four-Bar Linkages

A **bar** is a link containing only revolute pairs elements. **Four-bar linkages** are kinematic chains or mechanisms composed of four-bar mechanisms.

The bars of a linkage can be named according to their **size**.

s: the shortest length bar (*short*).

l: the longest bar (*large*).

D. Martins and E. H. Murai, *Mechanisms–Synthesis and Analysis with Applications in Robotics*, Mechanisms and Machine Science 183,
https://doi.org/10.1007/978-3-031-95187-9_3

p, q: the two remaining bars, regardless of which of the two is the longest or shortest.

In a four-bar mechanism, the bars can be named according to their **function**.

imput bar (*crank*), a: bar in which the motion or action input occurs. Often the input is made by inserting a motor into the kinematic pair that connects the input bar with the fixed bar.
coupler bar, b: bar that connects the input and output bars.
output bar, c: bar where the motion or action output occurs.
fixed bar, d: bar to which an absolute referential has been associated.

In Fig. 3.1 a four-bar mechanism is shown with the abbreviations referring to the function of each link.

For the mechanism in Fig. 3.1, if the sizes are $a = 7$ length units ($l.u.$), $b = 12\,l.u.$, $c = 15\,l.u.$ and $d = 11\,l.u.$, then

$$s = \min\{a, b, c, d\} = a = 7\,l.u.$$

$$l = \max\{a, b, c, d\} = c = 15\,l.u.$$

$$p, q = \{a, b, c, d\} - \{s, l\} \quad \rightarrow \quad p = b = 12\,l.u. \quad q = d = 11\,l.u.$$

A **point of interest** is a point belonging to a link, and the kinematics of this point is relevant to the mechanism design or analysis. For example, Fig. 1.9 shows a kinematic chain used for vehicle suspension. In a simpler analysis, disregarding tire deformation, the contact point between tire and track does not move relative to link 1. Thus, this point can be considered as belonging to link 1. The same occurs with link 2 and its respective tire-track contact point. In vehicle suspension, it is interesting to

Fig. 3.1 Four-bar mechanism

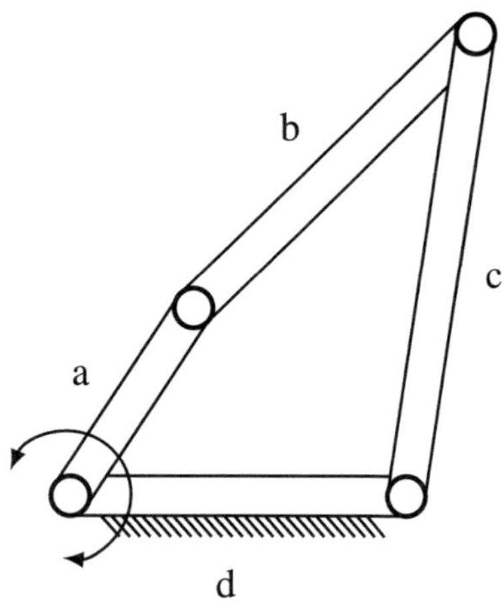

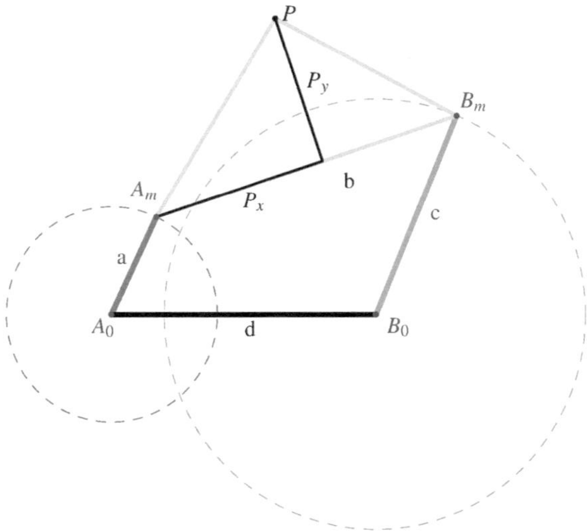

Fig. 3.2 Point of interest curve

study the kinematics of the contact point between tire and track as the suspension moves, so the contact point between tire and track is a point of interest.

A point of interest fixed on the coupler bar is called **coupler point**. The coupler point is widely used to do many different types of trajectories. An example of a coupler point trajectory is shown in Fig. 3.2. Changing the dimensions of the four-bar mechanisms (a, b, c and d) and the positioning of the point of interest on the coupler (p_x and p_y) changes the coupler point trajectory. Thus, the coupler point can generate curves of up to sixth order. Examples of coupler point curves are seen in Fig. 3.3. More content regarding coupler point curves can be seen in [1, 3].

In a four-bar mechanism, the input and output bars can be classified according to their **motion**.

Crank : bar capable of doing a complete rotation.
Rocker : bar that oscillates, being unable to do a full rotation.

A crank-type bar and a rocker-type bar can be seen in Fig. 3.4a, b, respectively.

The kinematics of mechanisms is defined from the kinematic pairs positioning, no matter the shape of the links. Thus, the mechanisms shown in Fig. 3.5 have the same kinematics, because the kinematic pairs are located in the same position for both mechanisms. In this way, the position of the kinematic pair is more relevant than the link itself. This characteristic is usually summarized in the humorous phrase "mechanisms are kinematic pairs connected by links, not links connected by kinematic pairs".

The **working mode** of a four-bar linkage (or simply a quadrilateral) is the way its links are assembled. When assembling a four-bar mechanism with a given angle to the input bar, the coupler and output bars can be assembled in two different ways. Each of these modes is said to be a working mode. In 3.6 the two working modes of a

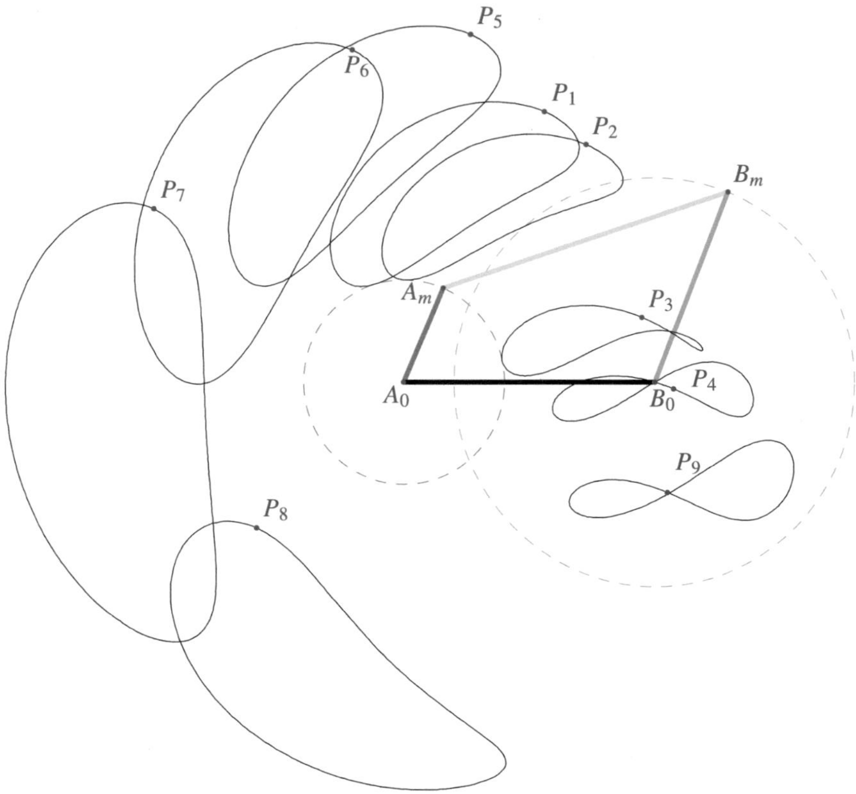

Fig. 3.3 Coupler point curves

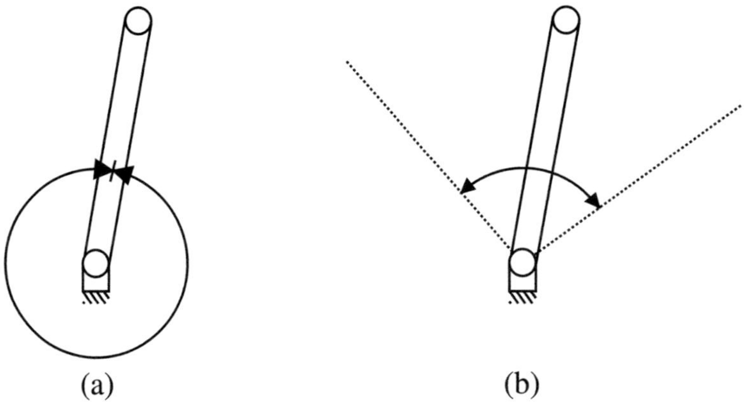

Fig. 3.4 Classification of a bar. **a** Crank. **b** Rocker

four-bar mechanism are shown. Some literature calls the working mode in Fig. 3.6a elbow up and that in Fig. 3.6b elbow down.

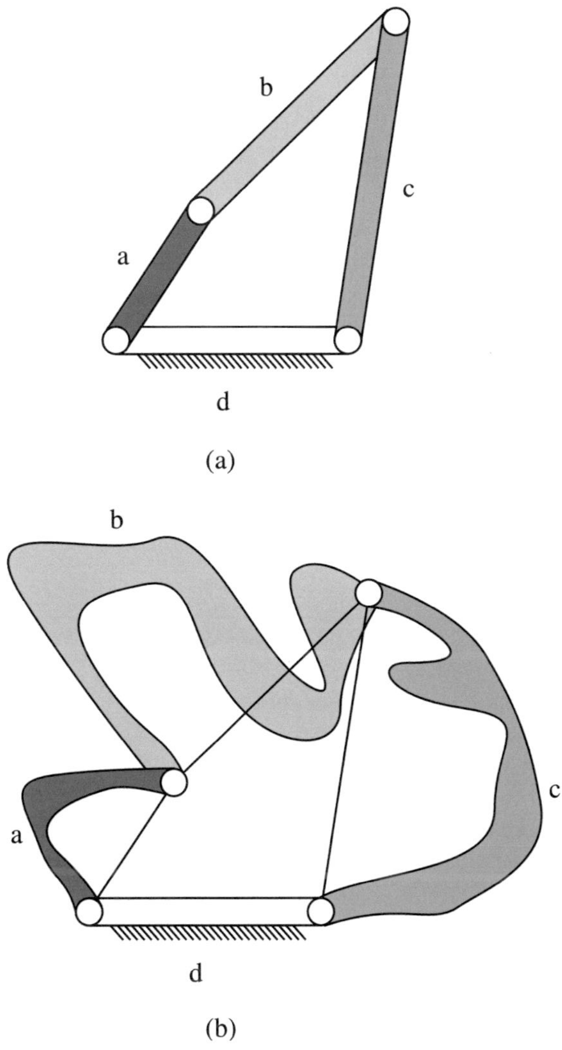

Fig. 3.5 Four-bar mechanism. **a** Straight links. **b** Complexly shaped links

3.2 Grashof Criterion

A linkage, whether a chain or a mechanism, can be classified into Grashof and non-Grashof.[1] In a Grashof linkage, the revolute pairs of the smaller link (s) are capable of doing complete revolution. A quadrilateral is said to be **Grashof** if it satisfies the following condition.

[1] Franz Grashof (1826–1893), German engineer known for his works in both machine design and thermodynamics.

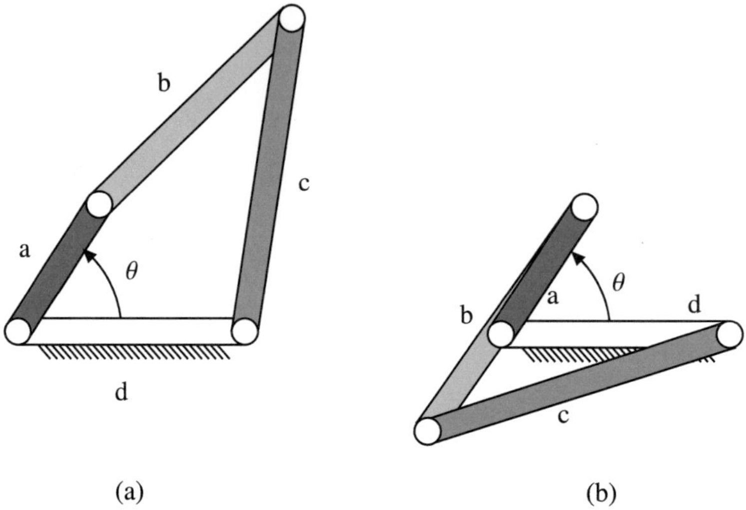

Fig. 3.6 Working modes of a four-bar mechanism

$$s + l \leq p + q. \tag{3.1}$$

If the quadrilateral does not satisfy such a condition, it is said not to be Grashof. In the boundary condition, when $s + l = p + q$, the four revolute pairs can align. In this case, the quadrilateral can change its working mode. In Fig. 3.7a, the quadrilateral is shown approaching alignment, in the Fig. 3.7b the quadrilateral is with all kinematic pairs aligned. The Fig. 3.7c shows the quadrilateral after alignment, and there are two possible working modes, one represented by the filled line and the other by the dashed line.

The setting $s + l = p + q$ should be avoided, because it is not certain which mode the mechanism will be in after alignment. If such a configuration is used, a limit switch must be inserted in order to prevent the alignment of the kinematic pairs of the mechanism.

Finally, in the position of Fig. 3.7b a special geometric condition occurs: the alignment of all kinematic pairs. As stated in Sect. 2.4, a differentiated behavior of the mechanism is expected in these special conditions. In fact, in this position the Jacobian matrix of the mechanism presents a zero line, which implies zero determinant, that is, a singularity problem. Therefore, just at this instant of time an additional actuator is required. This additional actuator would determine which working mode the mechanism would follow.

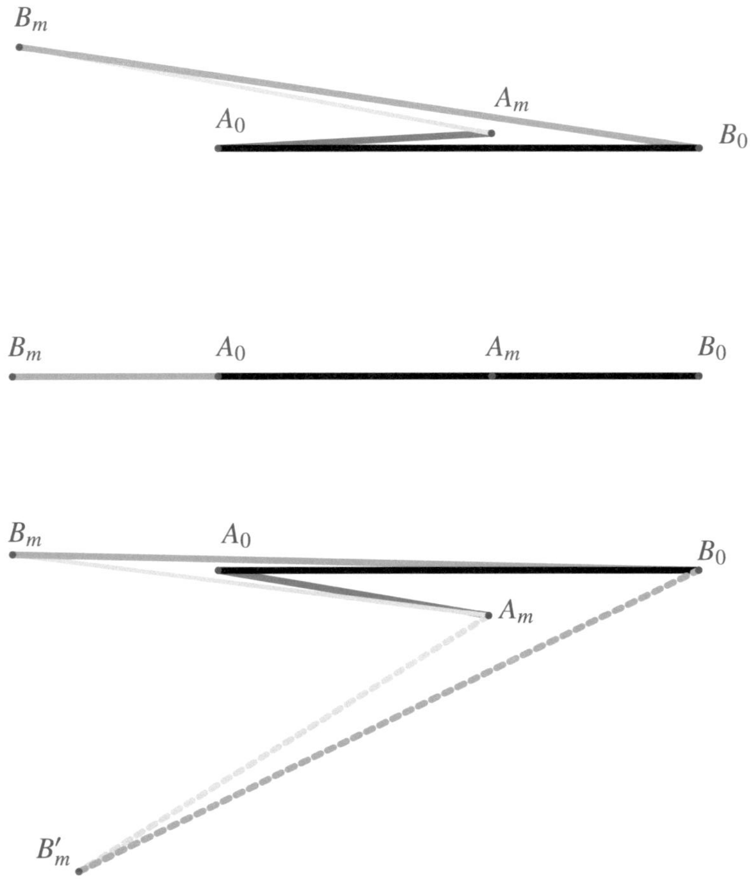

Fig. 3.7 Working modes. **a** Quadrilateral approaching alignment. **b** Quadrilateral aligned. **c** Quadrilateral leaving alignment

3.2.1 *Types of Grashof Quadrilaterals*

Grashof quadrilaterals are classified according to the input and output links motion, i.e. crank or rocker. A Grashof quadrilateral is a double-crank type if the smallest link is the fixed link. If the smallest link is the opposite to the fixed link, the mechanism will be a double-rocker. If the smallest link is the input link, then it will be a crank-rocker. If the smallest link is the output link, then the mechanism will be a rocker-crank. Thus, a Grashof quadrilateral can be classified into four types.

Double-crank: input bar and output bar are cranks. It occurs when $s \leftrightarrow d$. See example in Fig. 3.8a.

Crank-rocker: input bar is crank and output bar is rocker. It occurs when $s \leftrightarrow a$. See example in Fig. 3.8b.

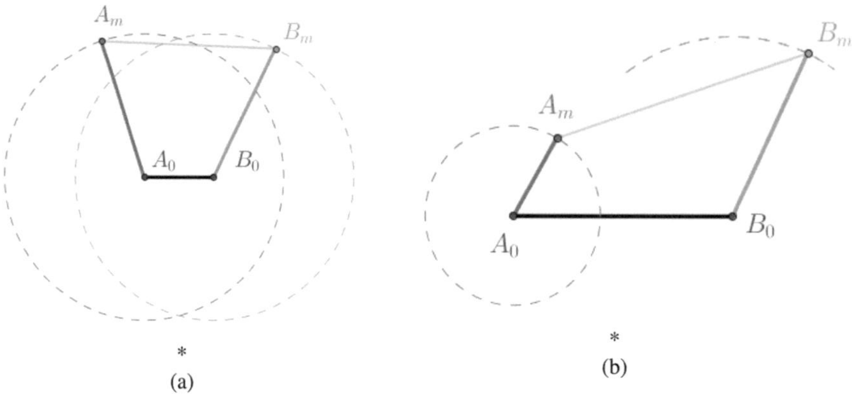

Fig. 3.8 Types of Grashof quadrilaterals. **a** Double-crank. **b** Crank-rocker

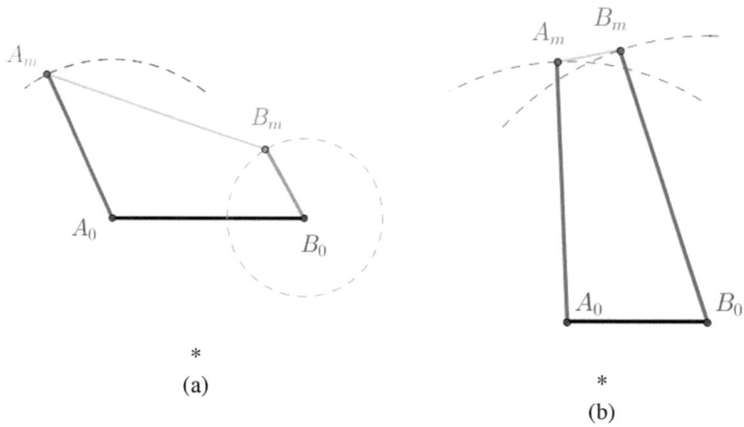

Fig. 3.9 Types of Grashof quadrilaterals. **a** Crank-rocker. **b** Double-rocker

Rocker-crank: input bar is rocker and output bar is crank. It occurs when $s \leftrightarrow c$. See example in Fig. 3.9a.

Double-rocker: input bar and output bar are rockers. It occurs when $s \leftrightarrow b$. See example in Fig. 3.9b.

Windshield wiper mechanism for vehicles

The function of a windshield wiper is to remove water and dirt from the windshield. There are several types of wipers, where the most common are: the *tandem* wiper, shown in Fig. 3.10a, the opposed wiper, shown in Fig. 3.10b, and the single wiper, shown in Fig. 3.10c. There is also the controlled single wiper, used e.g. on the Mercedes-Benz to increase the area of the windshield to be wiped. This last type is shown in Fig. 3.10d.

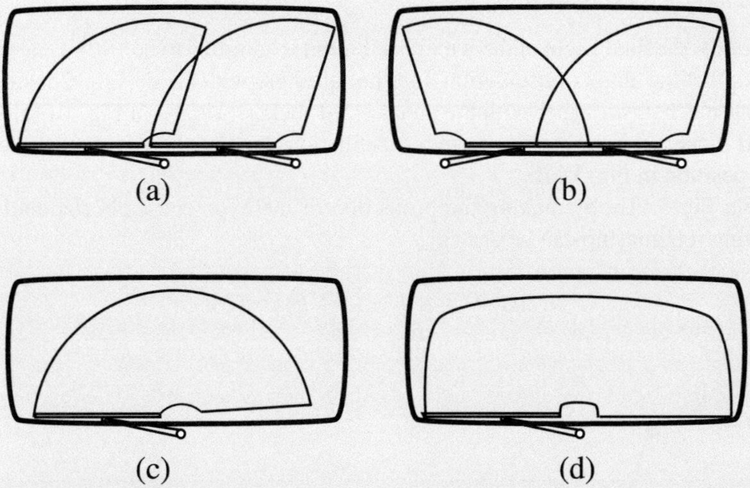

Fig. 3.10 Wiper types

Often, *tandem* type wipers use a Watt or a Stephenson mechanism. Such mechanisms can be thought of as two coupled four-bar mechanisms. One of these four-bar mechanisms is used to convert the rotational motion of the motor into the oscillating motion of the vanes (thus being a crank-rocker). Another four-bar mechanism is used to synchronize the two blades motion.

Design a single four-bar mechanism capable of actuating a windshield wiper of the *tandem* type from a rotary input. The vanes should rotate approximately 110°. What are the advantages and disadvantages of this new system over traditional ones that use the Watt or Stephenson kinematic chain?

Solution:
A windshield wiper of the *tandem* type has two wiper blade holders that oscillate. Thus, to use a single four-bar mechanism with two oscillating links, that four-bar mechanism must be a double-rocker.

If the actuator is to be a motor, then one of the links of the four-bar mechanism must be able to do a rotation. Therefore, the four-bar mechanism must be a double-rocker Grashof.

In order for both blades to do the same oscillation of approximately 110°, the four-bar mechanism must be symmetrical. Thus, the two rockers, having the same size, will also have the same oscillation range.

Thus, the design requirements are as follows:

1. Grashof $\rightarrow s + l \leq p + q$.
2. double-rocker $\rightarrow$ smallest link must be the opposite to the fixed link ($s \leftrightarrow b$).
3. Same size rockers $\rightarrow a = c$.

Initially, the final angle of the wiper oscillation is admitted to be 140°. Later on we will see that the angle of oscillation does not interfere with the design. Consider the four-bar mechanism wiper with the equal-sized rockers, shown in Fig. 3.11a. When the left rocker is at the starting point of oscillation, the four-bar mechanism will be at the position in Fig. 3.11b.

From Fig. 3.11b, by making the projection of the hypotenuses horizontally, the following relationship can be drawn:

$$a\cos(40^\circ) + d = (a + b)\cos(\beta). \tag{3.2}$$

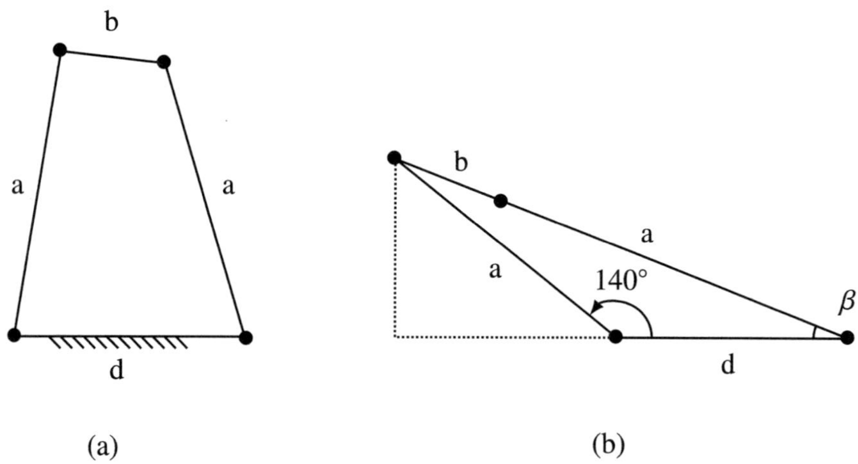

Fig. 3.11 Proposed windshield wiper. **a** Intermediate position. **b** Position where the θ angle assumes its largest value, 140°

One can determine β by using the law of sines:

$$\frac{a + b}{\sin(140^\circ)} = \frac{a}{\sin(\beta)}$$

$$\beta = \arcsin\left(\frac{a\sin(140^\circ)}{b + a}\right).$$

Substituting β into Eq. 3.2, one has

$$a\cos(40^\circ) + d = (a+b)\cos\left(\arcsin\left(\frac{a\sin(140^\circ)}{b+a}\right)\right).$$

To make the solution easier, we can admit the value of $a = 1\,l.u.$ Later on we will see that this choice does not influence the design either. Rearranging the terms, we have d as a function of b

$$d = (b+1)\cos\left(\arcsin\left(\frac{\sin(140^\circ)}{b+1}\right)\right) - \cos(40^\circ) \tag{3.3}$$

Choosing $b = 0.4\,l.u.$, the Eq. 3.3 results in $d = 0.477\,l.u.$ Calculating the value of β, one has $\beta \approx 27^\circ$. Thus, the rockers will oscillate approximately from 27° to 140°, i.e. 113° of oscillation, close to the desired value of 110°. It can be seen that in Fig. 3.11b the right-hand rocker can still rotate a bit more counterclockwise. Thus, the value of β can still be reduced. For example purposes, we will approximate $\beta \approx 27^\circ$. For convenience, one can round the value of d to $d = 0.5\,l.u.$. This gives the size of all bars in the mechanism.

Notes:

1. The resulting mechanism satisfies the design requirement of being Grashof, because with $a = 1\,l.u.$, $b = 0.4\,l.u.$, $c = 1\,l.u.$ and $d = 0.5\,l.u.$ one has that $s + l \leq p + q \rightarrow 0.4 + 1 \leq 0.5 + 1$.
2. The resulting mechanism satisfies the requirement of being double-rocker, because the smallest link is the opposite link to the fixed link.
3. We adopted $a = 1\,l.u.$ in order to normalize the dimensions. However, if other dimensions are desired one can multiply all sizes by the same factor. For example, if the fixed link is desired to be 60 cm long, the two rockers will be 120 cm long and the coupler link will be 48 cm long. Thus, the choice of $a = 1\,l.u.$ does not affect the design development.
4. It was adopted that the final angle of oscillation of the rockers was 140°. If another end angle is desired, for example 180°, it is sufficient that the vane has an additional angle of 40° in order to compensate for this difference. An example is seen in Fig. 3.12, in which the vane has an additional angle to the rocker.
5. To drive the mechanism, simply insert the rotation actuator (e.g. motor) into the kinematic pair that connects the coupler to the rocker (either rocker).

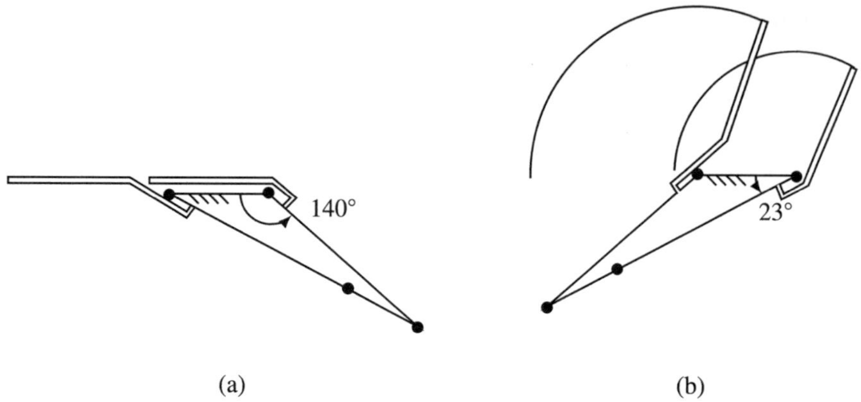

Fig. 3.12 Proposed windshield wiper

Advantages and Disadvantages

+ Fewer links and kinematic pairs, resulting in a mechanism with fewer components, simpler manufacturing and assembly, possibly with reduced cost.
– Less compact mechanism, requiring large space in the vehicle, motor is in a kinematic pair that does not belong to the fixed link, thus, when driving the mechanism, the motor moves along with the coupler link, increasing the system inertia.

3.2.2 *Types of Non-Grashof Quadrilaterals*

Non-Grashof quadrilaterals do not have kinematic pairs that do full rotation. Therefore, non-Grashof quadrilaterals are always double-rockers. Non-Grashof quadrilaterals can be classified according to the intersection of the rockers with the fixed link into two types.

Internal: rocker intersects with the fixed link from the internal side of the mechanism. See Fig. 3.13a.
External: rocker intersects with the fixed link from the external side of the mechanism. See Fig. 3.13b.

Thus, if the input rocker is internal and the output rocker is external, the mechanism is a *double internal-external rocker*. Therefore, a double-rocker can be classified as internal-internal, internal-external, external-internal, and external-external. Table 3.1 shows the classification of non-Grashof quadrilaterals.

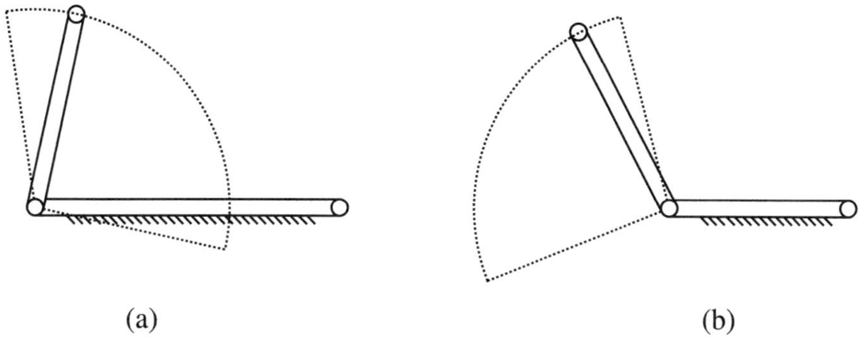

Fig. 3.13 Rocker classification. **a** Internal. **b** External

Table 3.1 Classification of non-Grashof quadrilaterals

		Output	
		Internal	External
Input	Internal	Internal-internal	Internal-external
	External	External-internal	External-external

3.3 Transmission Quality and Transmission Angle

Pressure angle (δ) is the angle between the *force* applied to a point and the *velocity* of the point. A simple example of pressure angle is the angle to push a block along a horizontal path, See Fig. 3.14.

For linkages, pressure angle is the angle between the force the coupler exerts on the output bar and the velocity of the application point on the output bar. Figure 3.15 shows the pressure angle for a four-bar mechanism.

The **transmission quality** is an indicator of the effectiveness of the mechanism with respect to its geometry. Transmission quality can be determined by the **transmission angle** (μ) between the *coupler* and the *output bar*. Figure 3.15 shows the transmission quality for a linkage.

Fig. 3.14 Pushing a block along a horizontal path with pressure angle (δ)

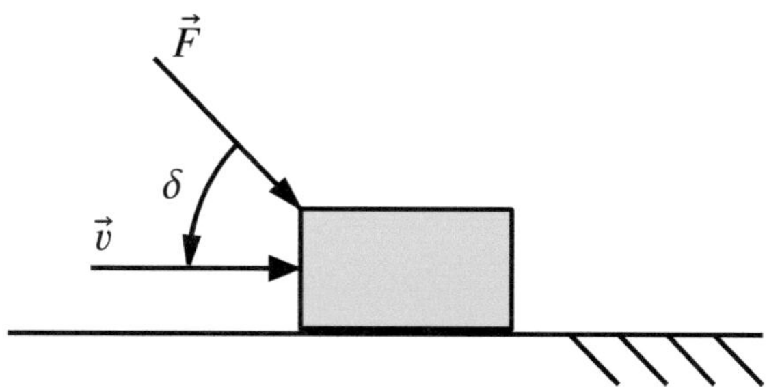

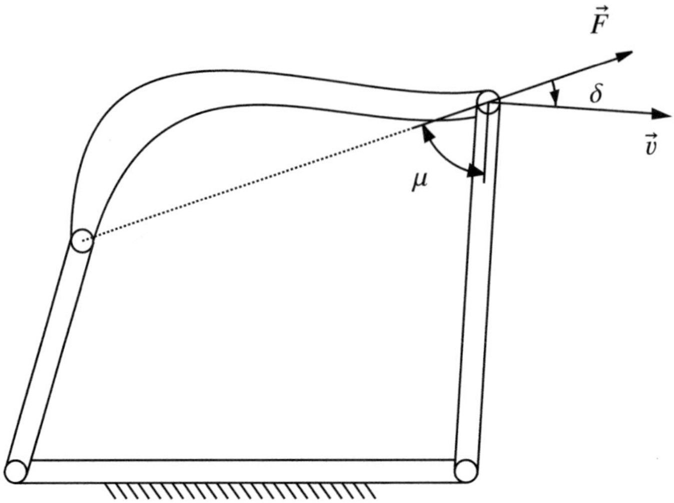

Fig. 3.15 Transmission angle μ and pressure angle δ

The optimal value for the pressure angle is when $\delta = 0°$. In this case, the whole force coming from the coupler is integrally used to move the output bar. When $\delta = 0°$, the transmission angle is 90°. In mechanism design, the recommended range for the pressure angle, according to [2], is

$$-50° \leq \delta \leq 50°$$

and the recommended range for the transmission angle is

$$40° \leq \mu \leq 140°.$$

3.3.1 Determining the Transmission Angle

A linkage can be divided into two triangles, as indicated in Fig. 3.16. In Fig. 3.16, θ is the input angle or drive angle, μ is the transmission angle, and e is the length of the diagonal of the quadrilateral.

Applying the law of cosines to both triangles, one has

$$e^2 = a^2 + d^2 - 2ad\cos(\theta) \tag{3.4}$$

$$e^2 = b^2 + c^2 - 2bc\cos(\mu). \tag{3.5}$$

Subtracting Eq. 3.5 from Eq. 3.4, one has

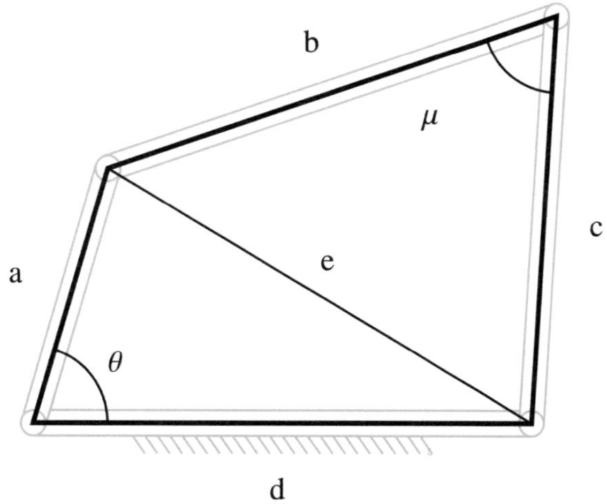

Fig. 3.16 Angles for determining the transmission angle

$$0 = a^2 + d^2 - b^2 - c^2 - 2ad\cos(\theta) + 2bc\cos(\mu).$$

Isolating $\cos(\mu)$, we have

$$\cos(\mu) = \frac{b^2 + c^2 - a^2 - d^2}{2bc} + \frac{2ad\cos(\theta)}{2bc}. \tag{3.6}$$

Defining A and B as

$$A = \frac{b^2 + c^2 - a^2 - d^2}{2bc} \tag{3.7}$$

$$B = \frac{ad}{bc} \tag{3.8}$$

One can rewrite Eq. 3.6 as

$$\mu = \arccos(A + B\cos(\theta)). \tag{3.9}$$

An important feature and one that will be used later is that B is always positive, because a, b, c and d are lengths, which must always be positive.

Determining the transmission angle graph

Determine the transmission angle graph as a function of input angle for a quadrilateral with the following measurements:

- $a = 4$
- $b = 2$

- $c = 4$
- $d = 5$

Solution:

Determining the values of A and B:

$$A = \frac{b^2 + c^2 - a^2 - d^2}{2bc} = \frac{2^2 + 4^2 - 4^2 - 5^2}{2 * 2 * 4} = A = \frac{-21}{16} = -1.3125$$

$$B = \frac{ad}{bc} = \frac{4 * 5}{2 * 4} = \frac{20}{8} = 2.5$$

The transmission angle function becomes

$$\mu = \arccos(-1.3125 + 2.5\cos(\theta))$$

Varying the input angle value gives the corresponding transmission qualities. One should be careful and analyze which type of four-bar mechanism it is. Note that this mechanism satisfies Grashof's criterion, given by $s + l \leq p + q = 2 + 5 \leq 4 + 4$. Since the smallest link is the link opposite to the fixed link (link b), then the mechanism is a double- rocker. Therefore, θ will not vary from 0° to 360°. One must determine the maximum and minimum values reached by the input angle.

The limits for the input angle occur when the mechanism is arranged in the two ways shown in Fig. 3.17.

The input angle in both cases can be determined using the law of cosines. For Fig. 3.17a one has:

$$(c + b)^2 = a^2 + d^2 - 2ad\cos(\theta)$$
$$\theta = \arccos\left(\frac{a^2 + d^2 - (c + b)^2}{2ad}\right) \approx 82.8°.$$

For Fig. 3.17b one has:

$$(c - b)^2 = a^2 + d^2 - 2ad\cos(\theta)$$
$$\theta = \arccos\left(\frac{a^2 + d^2 - (c - b)^2}{2ad}\right) \approx 22.3°.$$

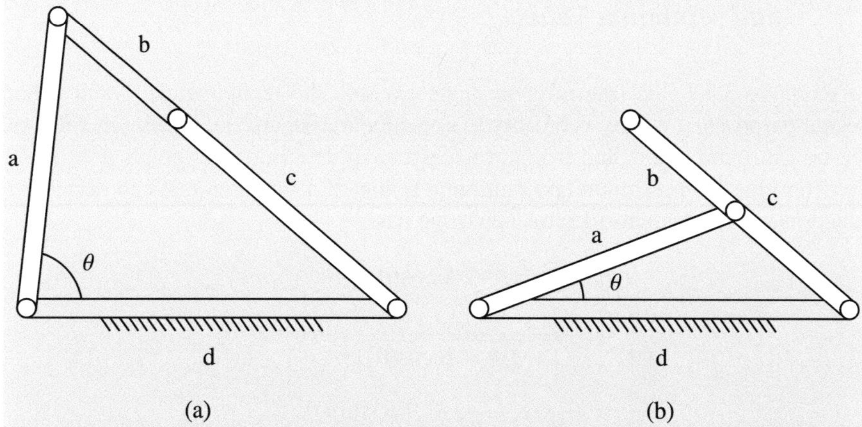

Fig. 3.17 Determining the limits for the input angle. **a** Largest θ value. **b** Lowest θ value

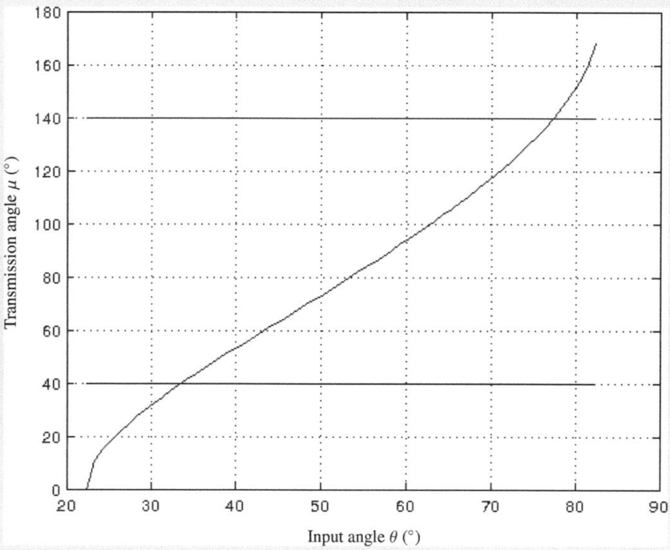

Fig. 3.18 Transmission angle for the double-rocker

Calculating the transmission angle, μ, by varying the input angle, θ, from 22.3° to 82.8°, the graph in Fig. 3.18 is obtained. The limits for the transmission angle recommended range ($40^\circ \leq 140^\circ$) are also shown in the graph. The script used to determine the graph in *Octave* [4] is shown in AppendixA.

3.3.2 *Determining the Transmission Angle Maximum and Minimum Values*

In Example 3.3.1, the transmission angle exceeds the recommended values. For design purposes, it is only necessary to know the maximum and minimum value of the transmission angle, and it is not necessary to determine the graph $\mu = f(\theta)$. To determine the maximum and minimum points of a function, one can derive the function and set it equals to zero. Thus, one has

$$\mu = \arccos(A + B\cos(\theta))$$
$$\frac{d\mu}{d\theta} = \frac{-1}{\sqrt{1-(A+B\cos(\theta))^2}}(-B\sin(\theta))$$
$$\frac{d\mu}{d\theta} = \frac{-1}{\sqrt{1-\cos^2(\mu)}}(-B\sin(\theta))$$
$$\frac{d\mu}{d\theta} = \frac{-1}{\sqrt{\sin^2(\mu)}}(-B\sin(\theta))$$
$$\frac{d\mu}{d\theta} = \frac{B\sin(\theta)}{\sin(\mu)}.$$

Equalizing to zero, we have

$$\frac{d\mu}{d\theta} = \frac{B\sin(\theta)}{\sin(\mu)} = 0 \rightarrow \theta = k\pi\ rad \qquad k \in [0; 1; 2; 3; ...[. \tag{3.10}$$

Equation 3.10 is verified when the input angle takes on integer multiples of π. Because of the range of recommended values for transmission angle, the term $\sin(\mu)$ in Eq. 3.10 will not be null.

Deriving again to determine whether it is maximum or minimum, we have

$$\frac{d\mu}{d\theta} = \frac{B\sin(\theta)}{\sin(\mu)}$$

$$\frac{d^2\mu}{d\theta^2} = \frac{B\cos(\theta)\sin(\mu) - B\sin(\theta)\cos(\mu)\frac{d\mu}{d\theta}}{\sin^2(\mu)}.$$

Since $\theta = k\pi$ with $k \in [0; 1; 2; 3; ...[$, it follows that $\sin(\theta) = 0$ for all stationary points (zero first derivative). Thus, the expression above reduces to

$$\frac{d^2\mu}{d\theta^2} = \frac{B\cos(\theta)\sin(\mu)}{\sin^2(\mu)}$$
$$\frac{d^2\mu}{d\theta^2} = \frac{B\cos(\theta)}{\sin(\mu)}.$$

Thus, one has

$$\frac{d^2\mu}{d\theta^2} = \frac{+B}{\sin(\mu)} \text{ para } \theta = 2k\pi$$
$$\frac{d^2\mu}{d\theta^2} = \frac{-B}{\sin(\mu)} \text{ para } \theta = (2k-1)\pi.$$

Note that B is always positive, and μ will be in a range of values for which $\sin(\mu)$ is positive. Thus, when $\theta = 0°$, the second derivative is positive, and one has the minimum point of the function. When $\theta = \pi\ rad$, the second derivative is negative, and one has the maximum point of the function.

Thus, substituting $\theta = 0$ and $\theta = \pi$ into Eq. 3.9 to obtain, respectively, the minimum and maximum transmission angle values, one has

$$\mu_{min} = \arccos(A + B) \tag{3.11}$$

$$\mu_{max} = \arccos(A - B). \tag{3.12}$$

Problems

3.1 Given a four-bar mechanism with dimensions $a = 1\ l.u.$, $b = 5\ l.u.$, $c = 3\ l.u.$ and $d = 4\ l.u.$, it is asked:

- with the aid of a compass and a ruler, draw the mechanism;
- determine μ_{max} and μ_{min}.

3.2 Determine the dimensions of a crank-rocker mechanism to meet the following design requirements:

- $b = 5\ l.u.$
- $c = 1\ l.u.$
- $\mu_{min} = 60°$
- $\mu_{max} = 120°$

3.3 A four-bar mechanism has the following dimensions $a = 1.8\ l.u.$, $b = 1\ l.u.$, $c = 2\ l.u.$ and $d = 1.2\ l.u.$ When in operation, the mechanism has its input angle ranging from 60° to 140°. Determine the highest and lowest value that the transmission angle reaches. Is this operating range within the recommended range?

3.4 Develop in GeoGebra[2]:

[2] Exercise for the reader to gain familiarity with using GeoGebra, an application that will be quite handy in later chapters.

- a non-Grashof linkage;
- a Grashof four-bar linkage.
  - make an inversion of the Grashof mechanism that will result in a crank-rocker mechanism.
  - make an inversion of the Grashof mechanism that will result in a double-crank mechanism.
  - make an inversion of the Grashof mechanism that will result in a rocker-crank mechanism.
  - make a reversal of the Grashof mechanism that will result in a double-rocker mechanism.

References

1. Hartenberg, R.S., Denavit, J.: Kinematic Synthesis of Linkages. McGraw-Hill, New York (1964)
2. Mabie, H.H., Ocvirk, F.W.: Mechanisms and Dynamic of Machinery. Wiley, New York (1963)
3. Hunt, K.H.: Kinematic Geometry of Mechanisms. Clarendon Press, Oxford (1978)
4. Eaton, J.W.: GNU Octave version 5.2.0 manual: a high-level interactive language for numerical computations (2020). Available at: https://www.gnu.org/software/octave/doc/v5.2.0/

Chapter 4
Structural Analysis

Abstract In this chapter, structural analysis is applied to kinematic chains to ensure the viability of mechanism configurations before engaging in detailed design. Key structural parameters such as variety, connectivity, and redundancy are introduced, providing insight into how these factors influence control and actuation points. The concept of variety, defined as the difference between global and minimal loop mobilities, is emphasized as a criterion for actuator selection. Matrix representations—degree of control, connectivity, and redundancy matrices—are explained and applied to various planar mechanisms. Isomorphism between graphs is discussed to identify structurally equivalent mechanisms and eliminate redundant solutions in mechanism enumeration. Numerous example figures illustrate complex kinematic chains and their structural decomposition. By performing structural analysis early in the design process, engineers can eliminate infeasible or inefficient mechanisms, streamlining the path to optimal solutions. This chapter forms a crucial link between conceptual design and functional mechanism synthesis.

Keywords Structural characteristics · Variety · Connectivity · Redundancy · Degree-of-control · Graphs · Isomorphism

This chapter covers the structural analysis of kinematic chains or mechanisms. The **structural analysis** is important because, when coupled with design requirements, it eliminates unfeasible results early in the mechanism design. It is recommended to review the structural characteristics and the concepts of kinematic pair connectivity and mobility seen in Sects. 1.1.7, 2.1 and 2.3 respectively.

D. Martins and E. H. Murai, *Mechanisms–Synthesis and Analysis with Applications in Robotics*, Mechanisms and Machine Science 183,
https://doi.org/10.1007/978-3-031-95187-9_4

4.1 Variety

The **variety** (V) of a mechanism is the difference between the mechanism mobility and the smallest mobility of the mechanism's closed subchains (M'). Variety is used to select those kinematic pairs that will receive actuators.

$$V = M - M'_{\min}. \tag{4.1}$$

Variety determination

Determine the variety of the kinematic chains with the expanded kinematic pairs shown in Fig. 4.1.

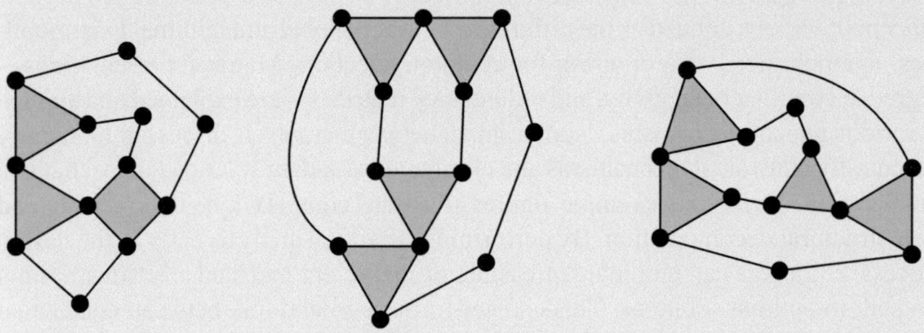

(a) (b) (c)

Fig. 4.1 Three planar kinematic chains with similar characteristics

Solution:
Figure 4.1**a kinematic chain**

The mobility of the kinematic chain will be calculated by the Eq. 2.4. The result is $M = (n - 1 - j)\lambda + j = (10 - 1 - 12)3 + 12 = 3$. By Eq. 2.5 one can determine the number of independent loops as $\nu = j - n + 1 = 12 - 10 + 1 = 3$. For each of the three independent loops (see Fig. 4.2a) the Eq. 2.4 is applied, obtaining the mobilities of each loop or subchain.

Subchain 1: $M' = (n - 1 - j)\lambda + j = (4 - 1 - 4)3 + 4 = 1.$
Subchain 2: $M' = (n - 1 - j)\lambda + j = (5 - 1 - 5)3 + 5 = 2.$
Subchain 3: $M' = (n - 1 - j)\lambda + j = (7 - 1 - 7)3 + 7 = 4.$

The mobility of each subchain is identified in Fig. 4.2a. Using the Euler equation, the number of independent loops can be calculated. However, caution must be exercised when identifying the loops. In some kinematic chains there may exist some loop, not evident in a first analysis, but with a lower M' mobility therefore affecting the variety of the kinematic chain.

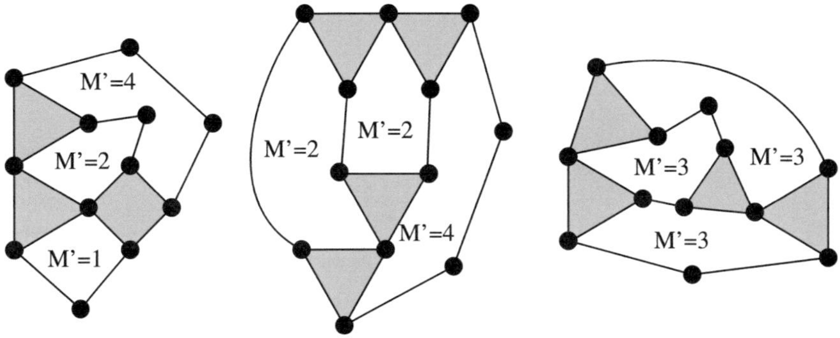

(a) (b) (c)

Fig. 4.2 Planar kinematic chains

The lowest subchain mobility is given by $M' = 1$. The variety is calculated by Eq. 4.1

$$V = M - M'_{\min} = 3 - 1 = 2.$$

Since the mobility of the kinematic chain is 3, one must insert actuators in three kinematic pairs. One can insert the first actuator into any kinematic pair, however, the last V actuators must be inserted with caution. In this example, the last two actuators must be positioned with caution. This can be seen by looking at subchain 1. If we insert the first actuator into any kinematic pair in this subchain, the other two actuators cannot be placed in pairs in the same subchain. If the actuators are inserted in such pairs, we get a subchain with mobility 1 but with two actuators, resulting in conflicting actuators.

Figure 4.1**b kinematic chain**

One can determine the kinematic chain mobility with Eq.2.4. The result is $M = (n - 1 - j)\lambda + j = (10 - 12 - 1)3 + 12 = 3$. By Eq. 2.5 we calculate the number of independent loops as $\nu = j - n + 1 = 12 - 10 + 1 = 3$. For each of the three independent loops the Eq. 2.4 is applied, obtaining the mobilities of each loop or subchain.

Subchain 1: $M' = (n - 1 - j)\lambda + j = (5 - 1 - 5)3 + 5 = 2.$
Subchain 2: $M' = (n - 1 - j)\lambda + j = (5 - 1 - 5)3 + 5 = 2.$
Subchain 3: $M' = (n - 1 - j)\lambda + j = (7 - 1 - 7)3 + 7 = 4.$

The mobility of each subchain is identified in Fig. 4.2b.

The smallest subchain mobility is given by $M' = 2$. The variety is calculated by Eq. 4.1

$$V = M - M'_{\min} = 3 - 2 = 1.$$

In this example, the last actuator must be positioned with caution. This means that one can insert two actuators in any positions and there will be no conflict; however, the last actuator should be positioned with caution to avoid conflicts between actuators.

Figure 4.1**c kinematic chain**

We determine the kinematic chain mobility by Eq. 2.4. The result is $M = (n - 1 - j)\lambda + j = (10 - 12 - 1)3 + 12 = 3$. By Eq. 2.5 the number of independent loops is $\nu = j - n + 1 = 12 - 10 + 1 = 3$. For each of the three independent loops the Eq. 2.4 is applied, obtaining the mobilities of each loop or subchain.

Subchain 1: $M' = (n - 1 - j)\lambda + j = (6 - 1 - 6)3 + 6 = 3.$
Subchain 2: $M' = (n - 1 - j)\lambda + j = (6 - 1 - 6)3 + 6 = 3.$
Subchain 3: $M' = (n - 1 - j)\lambda + j = (6 - 1 - 6)3 + 6 = 3.$

The mobility of each subchain is identified in Fig. 4.2c.

The smallest subchain mobility is given by $M' = 3$. The variety is calculated by Eq. 4.1

$$V = M - M'_{\text{min}} = 3 - 3 = 0.$$

In this example, since the kinematic chain has zero variety, one can insert the three actuators into any kinematic pairs that no conflict will occur between the actuators.

For more information regarding variety, see [1, 2].

4.2 Degree of Control, Connectivity and Redundancy

Degree of control, connectivity and redundancy are properties measured **between two links of a kinematic chain or mechanism** (usually with all kinematic pairs expanded). Thus, such properties are always cited as "degree of control between the i and j links" or "degree of control of the j link with respect to the i link". By analyzing the properties of all n links with respect to all n links, one has a $n \times n$ matrix. Also, as will be seen, analyzing such properties of the i link with respect to the j link or the j link with respect to the i link results in the same value, so such matrices should be symmetric. Thus, the three properties presented here are represented by symmetric $n \times n$ matrices.

An important remark for the reader is not to confuse the concepts of *kinematic pair connectivity* (f_i) and *connectivity* (C_{ij}). The former refers to the amount of degrees of freedom that a kinematic pair allows between the two links that the pair connects (see section 2.1), the latter refers to the amount of degrees of freedom between two links in a kinematic chain, as it will be seen. Another observation is that, since the properties are measured between links, it is not relevant whether the kinematic chain is fixed or not. That is, such properties are the same for a kinematic chain or for a mechanism, resulting in the same matrices in both chains and mechanisms.

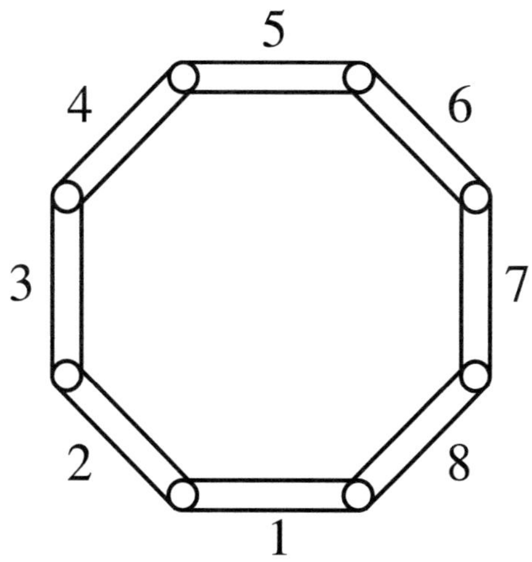

Fig. 4.3 Closed kinematic chain

The **degree of control** between the i and j links, represented by K_{ij}, is the **minimum** number of kinematic pairs that must be constrained in a kinematic chain for the j link to have no degree of freedom (motion) with respect to the i link. Note that the j link not moving relative to the i link is equivalent to the i link not moving relative to the j link, so $K_{ij} = K_{ji}$. Such equality implies symmetry in the K matrix, as it was mentioned in the preamble to this section.

Often the degree of control is defined as the minimum number of *actuators* required to be inserted into the chain so that the j link does not move relative to the i link. For this, it is considered that inserting an actuator controls the relative displacement between two parts, setting it to a value. Thus, a kinematic pair is restricted. For example, consider a chain consisting of two links connected by a revolute pair. If an actuator is inserted into the kinematic pair and a given constant angle is imposed, the links can no longer rotate relative to each other. Thus, by inserting an actuator and fixing its displacement, one degree of freedom is constrained.

Note that the largest possible value for K_{ij} is M, the mobility of the mechanism itself. This is because by restricting M degrees of freedom of the chain, the position of all links within it will be completely defined, including the i and j links.

Detemination of the degree of control matrix (K_{ij})

For the kinematic chain shown in Fig. 4.3, determine the degree of control matrix. All kinematic pairs are revolute.

Solution:

The degree of control between links i and j is the amount of actuators required to completely define the position of the j link relative to the i link. Considering $i = 1$ and varying j, one has:

j = 1 : no actuator is required, since link 1 does not move relative to link 1 $K_{11} = 0$.

j = 2 : it is sufficient to constrain the kinematic pair connecting link 1 to link 2. Thus, an actuator is needed between these two links $K_{12} = 1$.

j = 3 : just constrain the pairs connecting link 1 to link 2 and link 2 to link 3. Thus, two actuators are needed $K_{13} = 2$.

j = 4 : just constrain the pairs connecting link 1 to link 2, link 2 to link 3 and link 3 to link 4. Thus, three actuators are needed $K_{14} = 3$.

j = 5 : just constrain the pairs connecting link 1 to link 2, link 2 to link 3, link 3 to link 4 and link 4 to link 5. Thus, four actuators are needed $K_{15} = 4$.

j = 6 : just constrain the pairs connecting link 1 to link 8, link 8 to link 7 and link 7 to link 6 Thus, three actuators are needed $K_{16} = 3$.

j = 7 : just constrain the pairs connecting link 1 to link 8 and link 8 to link 7. Thus, two actuators are needed $K_{17} = 2$.

j = 8 : just constrain the pair that connects link 1 to link 8. Thus, an actuator is needed $K_{18} = 1$.

Thus, the line $i = 1$ in the matrix K becomes

$$K = \begin{bmatrix} 0 & 1 & 2 & 3 & 4 & 3 & 2 & 1 \\ - & - & - & - & - & - & - & - \\ - & - & - & - & - & - & - & - \\ - & - & - & - & - & - & - & - \\ - & - & - & - & - & - & - & - \\ - & - & - & - & - & - & - & - \\ - & - & - & - & - & - & - & - \\ - & - & - & - & - & - & - & - \end{bmatrix} \tag{4.2}$$

Repeating the analysis for i from 2 to 8 completes the K matrix.

$$K = \begin{bmatrix} 0 & 1 & 2 & 3 & 4 & 3 & 2 & 1 \\ 1 & 0 & 1 & 2 & 3 & 4 & 3 & 2 \\ 2 & 1 & 0 & 1 & 2 & 3 & 4 & 3 \\ 3 & 2 & 1 & 0 & 1 & 2 & 3 & 4 \\ 4 & 3 & 2 & 1 & 0 & 1 & 2 & 3 \\ 3 & 4 & 3 & 2 & 1 & 0 & 1 & 2 \\ 2 & 3 & 4 & 3 & 2 & 1 & 0 & 1 \\ 1 & 2 & 3 & 4 & 3 & 2 & 1 & 0 \end{bmatrix} \tag{4.3}$$

The **connectivity** between the i and j links, represented by C_{ij}, is the relative mobility between the i and j links. Thus, connectivity refers to the number of degrees of freedom existing between the i and j links of a kinematic chain.

The reader should recall the definition of degree of freedom, provided in the section 1.1.7. Thus, connectivity is the number of possible *independent* motions between i and j, so its value is limited to a maximum value λ. Also, the number of independent motions that j can do relative to i is the same that i can do relative to j, so $C_{ij} = C_{ji}$, which implies symmetry of the matrix C.

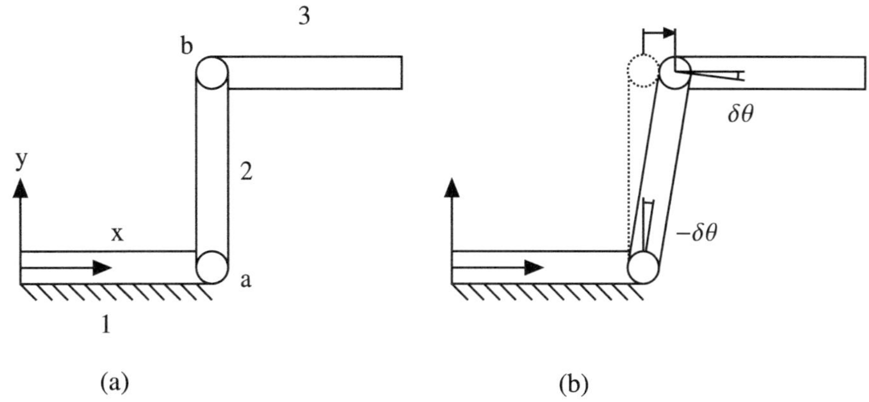

Fig. 4.4 Translational motion. **a** Before rotations. **b** After rotations

Even if a mechanism has only one type of kinematic pair, it is possible to create chains such that the linear combination of these kinematic pairs results in other possible motions. To be clearer, consider the mechanism in Fig. 4.4, which has only revolute pairs.

Link 2 can rotate relative to link 1, and link 3 can rotate relative to link 2. Although these two motions are of the same type, rotations, notice that link 3 relative to link 1 can make more motions than a simple rotation. This is due to the fact that the motions of the two kinematic pairs of rotation are combined, resulting in another independent motion.

By fixing link 1, one can rotate pair a by $-\delta\theta$ degrees. By rotating pair b of $\delta\theta$ degrees, link 3 will do a translation along x relative to link 1, as shown in Fig. 4.4b. Note that this motion is a pure translation, as the orientation of 3 relative to 1 remains unchanged.

The direction of translation varies with the position of the kinematic pairs, so the direction is always normal to the line that passes through the two kinematic pairs, as shown in Fig. 4.5a. With the mechanism being in the plane, the greatest possible connectivity between links 1 and 3 is $\lambda = 3$. Since link 3 has two independent motions relative to link 1, there must be a constraint between such links. The constrained motion is the translation of link 3 relative to link 1 along the straight line that passes through the two kinematic pairs, as shown in Fig. 4.5a.

In Fig. 4.5b another revolute pair is added, so that link 4 can translate in two different directions and also rotate relative to link 1. Thus link 4 has connectivity 3 with respect to link 1, and can do both translations and rotation in the plane. In the special position where the three kinematic pairs are aligned, the two translations become parallel and are dependent. Thus, in this aligned position, link 4 has connectivity 2 with respect to link 1.

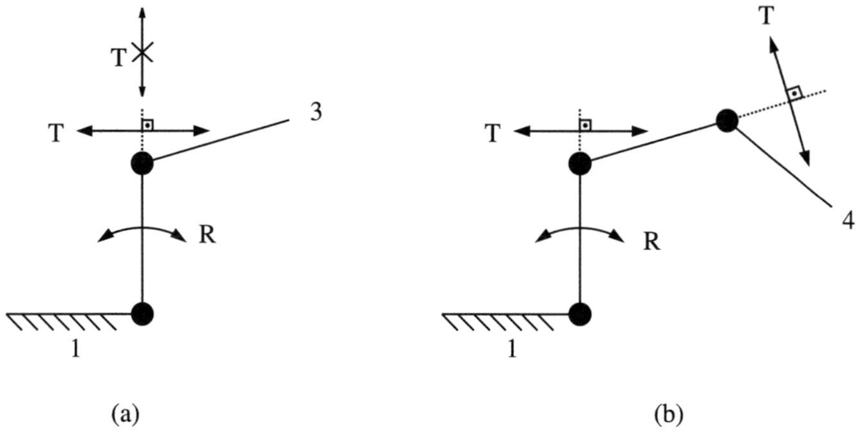

Fig. 4.5 Combination of revolute pairs. **a** One rotation and one translation. **b** One rotation and two translations

Determining the connectivity matrix

For the planar open kinematic chain in Fig. 4.6, determine the connectivity matrix. All kinematic pairs are revolute.

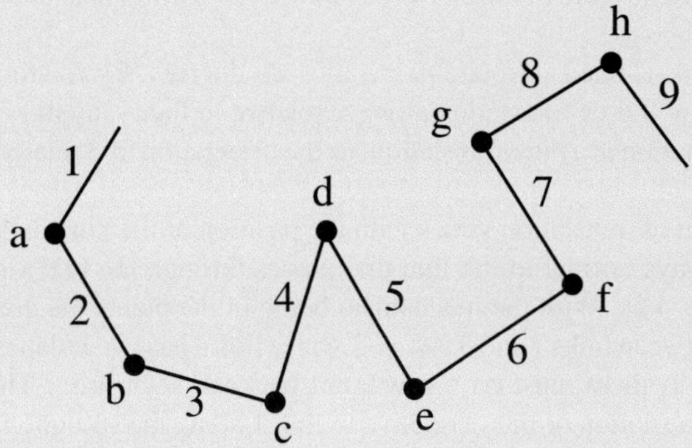

Fig. 4.6 Open planar kinematic chain

Solution:
Given $i = 1$ and varying j, one has:

j =1 : link 1 has no motion relative to link 1 $C_{11} = 0$.
j = 2 : link 2 can rotate relative to link 1. Thus, the number of independent motions is 1 $C_{12} = 1$.

j* = *3 : between links 1 and 3 there are two revolute pairs. The motions of these pairs can be combined to generate a rotation and a translation. Thus, the number of independent motions is 2 $C_{13} = 2$.

j* = *4 : between links 1 and 4 there are three revolute pairs. The motions of these pairs can be combined to generate one rotation and two translations. Thus, the number of independent motions is 3 $C_{14} = 3$.

j* = *5 : between links 1 and 5 there are four revolute pairs. The motions of these pairs can be combined to generate one rotation and two translations. Any motion between these two links can be described as a linear combination of the rotation and the two translations. Another way to arrive at the same result is to analyze the workspace. As stated in the statement, the chain is planar, so any relative motion between links must be contained within this space. Thus, like planar space, any motion can be written as a linear combination of one rotation and two translations. Therefore connectivity is said to be limited to a maximum value λ. Thus, the number of independent moves is 3 $C_{15} = 3$.

j* = *6*, *j* = *7*, *j* = *8* and *j* = *9 : as well as $j = 5$, $C_{16} = C_{17} = C_{18} = C_{19} = 3$.

Thus, the row $i = 1$ in the matrix C becomes

$$C = \begin{bmatrix} 0 & 1 & 2 & 3 & 3 & 3 & 3 & 3 & 3 \\ - & - & - & - & - & - & - & - & - \\ - & - & - & - & - & - & - & - & - \\ - & - & - & - & - & - & - & - & - \\ - & - & - & - & - & - & - & - & - \\ - & - & - & - & - & - & - & - & - \\ - & - & - & - & - & - & - & - & - \\ - & - & - & - & - & - & - & - & - \\ - & - & - & - & - & - & - & - & - \end{bmatrix} \tag{4.4}$$

Repeating the analysis for i from 2 through 9 completes the C matrix.

$$C = \begin{bmatrix} 0 & 1 & 2 & 3 & 3 & 3 & 3 & 3 & 3 \\ 1 & 0 & 1 & 2 & 3 & 3 & 3 & 3 & 3 \\ 2 & 1 & 0 & 1 & 2 & 3 & 3 & 3 & 3 \\ 3 & 2 & 1 & 0 & 1 & 2 & 3 & 3 & 3 \\ 3 & 3 & 2 & 1 & 0 & 1 & 2 & 3 & 3 \\ 3 & 3 & 3 & 2 & 1 & 0 & 1 & 2 & 3 \\ 3 & 3 & 3 & 3 & 2 & 1 & 0 & 1 & 2 \\ 3 & 3 & 3 & 3 & 3 & 2 & 1 & 0 & 1 \\ 3 & 3 & 3 & 3 & 3 & 3 & 2 & 1 & 0 \end{bmatrix} \tag{4.5}$$

Although the definitions of *degree of control* and *connectivity* are similar, the matrices K and C differ in the *ij* elements where $K_{ij} > \lambda$.

Comparison of the degree of control K_{ij} and connectivity matrices C_{ij}

For the planar kinematic chain shown in Fig. 4.3, determine the connectivity matrix. All kinematic pairs are revolute.

Solution:
The connectivity matrix of the chain is

$$C = \begin{bmatrix} 0\,1\,2\,3\,3\,3\,2\,1 \\ 1\,0\,1\,2\,3\,3\,3\,2 \\ 2\,1\,0\,1\,2\,3\,3\,3 \\ 3\,2\,1\,0\,1\,2\,3\,3 \\ 3\,3\,2\,1\,0\,1\,2\,3 \\ 3\,3\,3\,2\,1\,0\,1\,2 \\ 2\,3\,3\,3\,2\,1\,0\,1 \\ 1\,2\,3\,3\,3\,2\,1\,0 \end{bmatrix} \tag{4.6}$$

The control degree matrix, determined in Example 4.2, is

$$K = \begin{bmatrix} 0\,1\,2\,3\,4\,3\,2\,1 \\ 1\,0\,1\,2\,3\,4\,3\,2 \\ 2\,1\,0\,1\,2\,3\,4\,3 \\ 3\,2\,1\,0\,1\,2\,3\,4 \\ 4\,3\,2\,1\,0\,1\,2\,3 \\ 3\,4\,3\,2\,1\,0\,1\,2 \\ 2\,3\,4\,3\,2\,1\,0\,1 \\ 1\,2\,3\,4\,3\,2\,1\,0 \end{bmatrix} \tag{4.7}$$

Note that the matrices are similar, they only differ in the elements $K_{ij} = 4$, which become $C_{ij} = 3$ due to $\lambda = 3$. Thus, a practical way to determine such matrices is to start with the control degree matrix, as this is easier to determine intuitively. Then, to obtain the connectivity matrix one simply rewrites K by replacing the elements greater than λ with the value of λ.

The **redundancy** between the i and j links, represented by R_{ij}, is the difference between the degree of control and connectivity between the i and j links. Thus, the redundancy matrix can be obtained from the expression

$$R = K - C \tag{4.8}$$

When the redundancy between two i and j links is greater than zero ($R_{ij} \geq 1$), the positioning of the i link relative to the j link can be accomplished positioning the intermediate links in several different ways. When the R matrix has at least one positive element, the kinematic chain is said to be redundant or exhibit redundancy. A redundant kinematic chain can accomplish a task in several ways.

Redundancy matrix (R_{ij}) determination

For the planar robotic arm mechanism in Fig. 4.7, determine the redundancy matrix. All kinematic pairs are revolute.

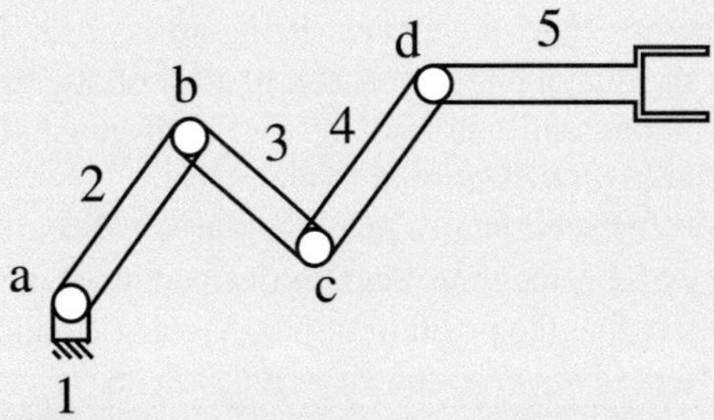

Fig. 4.7 Planar robotic arm

Solution:
The degree of control matrix is shown below

$$K = \begin{bmatrix} 0 & 1 & 2 & 3 & 4 \\ 1 & 0 & 1 & 2 & 3 \\ 2 & 1 & 0 & 1 & 2 \\ 3 & 2 & 1 & 0 & 1 \\ 4 & 3 & 2 & 1 & 0 \end{bmatrix} \tag{4.9}$$

The connectivity matrix is shown below

$$C = \begin{bmatrix} 0 & 1 & 2 & 3 & 3 \\ 1 & 0 & 1 & 2 & 3 \\ 2 & 1 & 0 & 1 & 2 \\ 3 & 2 & 1 & 0 & 1 \\ 3 & 3 & 2 & 1 & 0 \end{bmatrix} \tag{4.10}$$

The redundancy matrix is given by $R = K - C$

$$R = \begin{bmatrix} 0\,0\,0\,0\,1 \\ 0\,0\,0\,0\,0 \\ 0\,0\,0\,0\,0 \\ 0\,0\,0\,0\,0 \\ 1\,0\,0\,0\,0 \end{bmatrix} \tag{4.11}$$

Note that there is redundancy between links 1 and 5. Thus, one can completely define the position of 5 relative to 1 in several ways. Note that "completely defining the position" means restricting all degrees of freedom in the workspace. Since the statement says it is a planar robotic arm, it means that the workspace is planar, and one must restrict both translations in the plane and also rotation about an axis normal to the plane.

Figure 4.8 shows the robotic arm reaching a point P with a given angle θ, thus restricting x, y and θ, the three degrees of freedom of the plane. Note that one can completely define the position of link 5 (robotic hand) relative to link 1 (base) in two different ways, as shown in Fig. 4.8a, b.

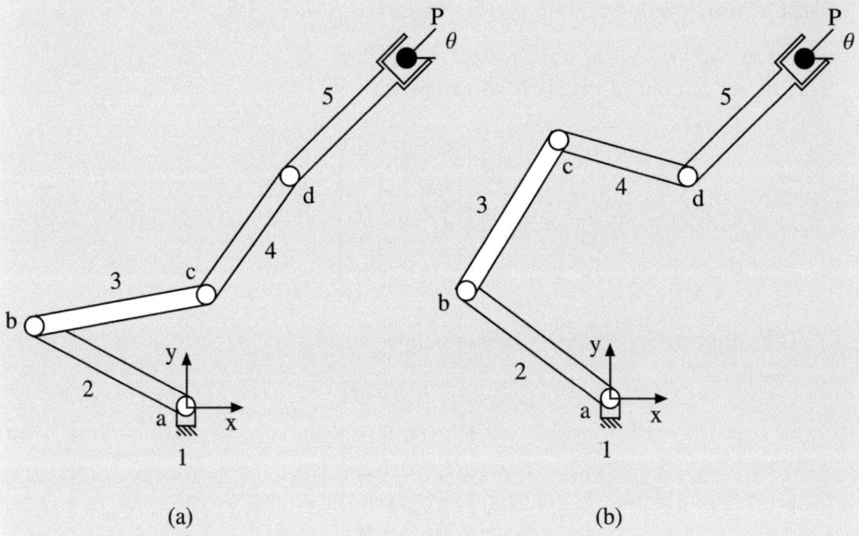

Fig. 4.8 Planar robotic arm

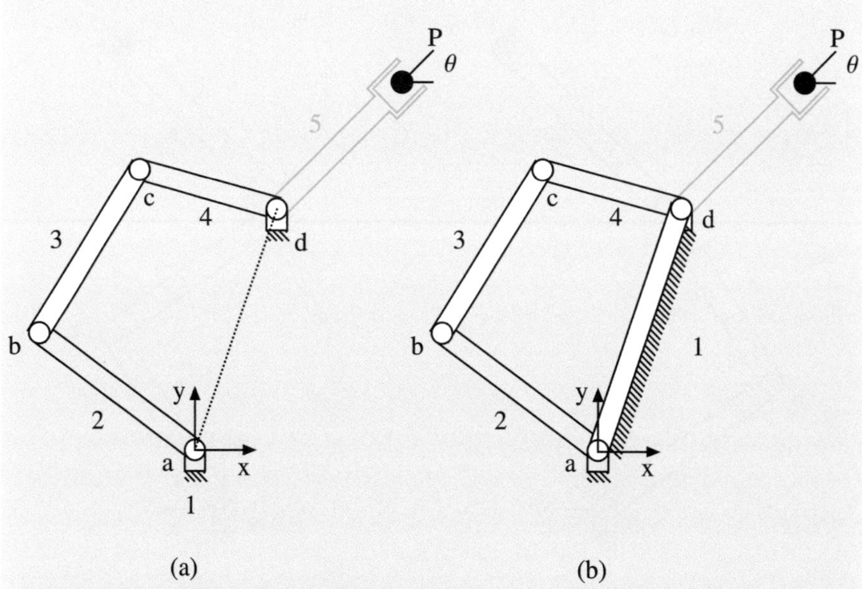

Fig. 4.9 planar robotic arm

In Fig. 4.8, the positioning the robotic hand at point P and with a given angle θ implies completely defining the position of link 5 with respect to the coordinate system. Any other position or angle of link 5 would not reach position P or orientation θ. Therefore, the position of the kinematic pair d is unique. Thus, one can design the mechanism shown in Fig. 4.9a. By analyzing the remaining links (see Fig. 4.9b), one has a mechanism of four links (links 1, 2, 3, and 4) and one loop. By Eq. 2.4, you have mobility 1, that is, one degree of freedom. Such a degree of freedom can take on infinite values, resulting in infinite modes for the robotic hand to reach the point P with angle θ. Thus, there are not only the two modes shown in Fig. 4.8, but there are infinite modes.

In Example 4.2 it was shown that when the redundancy between two links is 1, then one can do a task in infinitely many different ways. If between two links there is redundancy 0, it does not imply that there is only one way to do a task but rather finitely many ways to do a task. By eliminating link 5 from the robotic arm, the mechanism will no longer have redundancy. However, it will still be possible to reach a point P with a given angle θ in two different ways, as shown in Fig. 4.10.

It can be seen that the matrices K, C and R depend on how the links were numbered. Thus, **before determining such matrices, one must enumerate the links**. For more information about degree of control, connectivity and redundancy, see [1–4].

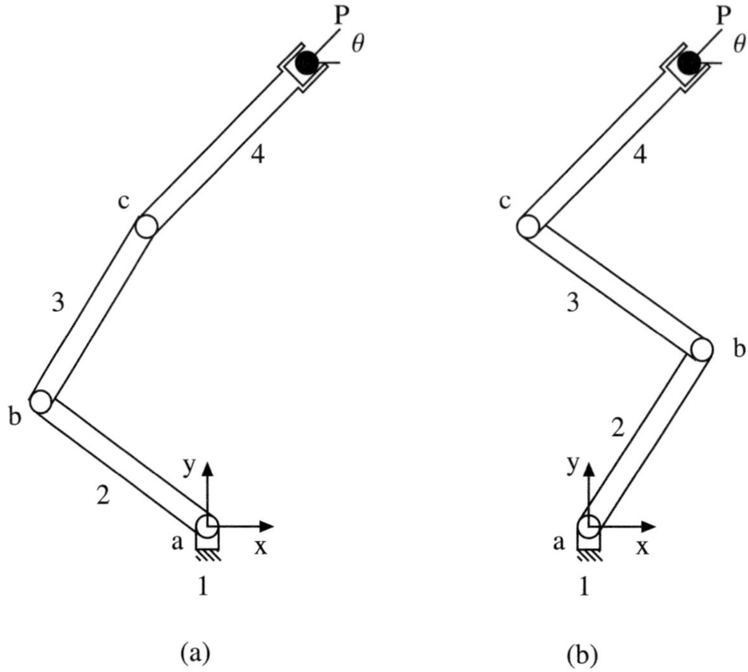

Fig. 4.10 Reduced planar robotic arm

4.3 Remembering Graphs

In section 1.1.6, the representation of kinematic chains and mechanisms using graphs was introduced. In this section, the representation via graphs will be looked at in more detail.

In graph representation, the links are represented by vertices and the kinematic pairs are represented by edges. Thus, a binary link will be represented by a vertex with two edges, a ternary link by a vertex with three edges, a quaternary link by a vertex with four edges and so on, as shown in Table 4.1.

It can be observed that between a graph and a kinematic chain there is a biunivocal relation, that is, given a kinematic chain, there is only one graph that represents it, and, given a graph, one can extract an unique kinematic chain. Graphs are independent of the dimensions of the edges and the positions of the vertices. Thus, one can change the appearance of the graph by moving its vertices, however, the connections between vertices will remain unchanged. This independence of dimensions is an interesting feature in graphs, because in structural analysis the dimensions are also not relevant (see *structure of a mechanism* in section 1.1.7). Thus, the structural characteristics of a chain can be studied through its graph.

One of the advantages of doing structural analysis through graphs is that there are more mathematical and computational tools for studying graphs than for study-

Table 4.1 Links and the representation via graphs

	Binary	Ternary	Quaternary
Link	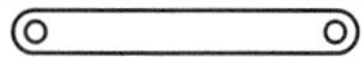	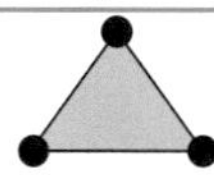	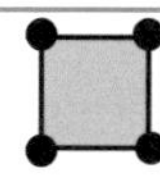
Graph		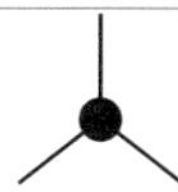	

Table 4.2 Bijective relationship between the links in Fig. 4.11a and 4.11b

Fig. 4.11a link	1	2	3	4	5	6	7	8
Fig. 4.11b link	2	4	6	5	3	1	7	8

ing kinematic chains. For example, one can represent a graph through its adjacency matrix and, from such matrix, obtain the matrices of degree of control, connectivity, and redundancy. Implementing a program to manipulate matrices is easier and faster than a program to directly analyze mechanisms, since the latter would involve image processing. Considering synthesis (a topic covered in Chap. 6), there are more promising tools for enumerating graphs than kinematic chains.

4.4 Isomorphism

Isomorphism is a condition in which two objects have the same shape. In the area of mechanism theory, two kinematic chains are said to be isomorphic if they have the same or equivalent structures. Two isomorphic kinematic chains differ only in the name given to their links and kinematic pairs. In this way, a biunivocal relationship can be established between the components of the two kinematic chains. Examples of isomorphic kinematic chains are shown in Fig. 4.11.

In Fig. 4.11 two kinematic chains are shown. A necessary but not sufficient condition for two kinematic chains to be isomorphic is that both have the same amount of links for each link type. Note that both the kinematic chain in Fig. 4.11a and the kinematic chain in Fig. 4.11b have four ternary links and four binary links. The sufficient condition for two kinematic chains to be isomorphic is that there exists a **bijector** function that correlates one link in one chain to another link in another chain. That is, for each link in the chain in Fig. 4.11a there is a single corresponding link in Fig. 4.11b. This bijection relationship for the chains in Fig. 4.11 is shown in Table 4.2.

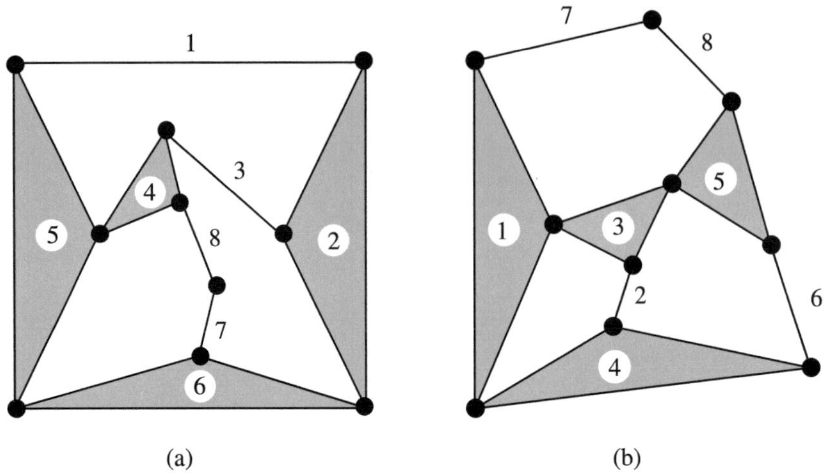

Fig. 4.11 Isomorphism between kinematic chains

If the kinematic chains in Fig. 4.11 are isomorphic, then their structural features must be the same.

Properties of isomorphic kinematic chains

For the two kinematic chains in Fig. 4.12, determine the degree of control, connectivity, redundancy matrices, and their graphs.

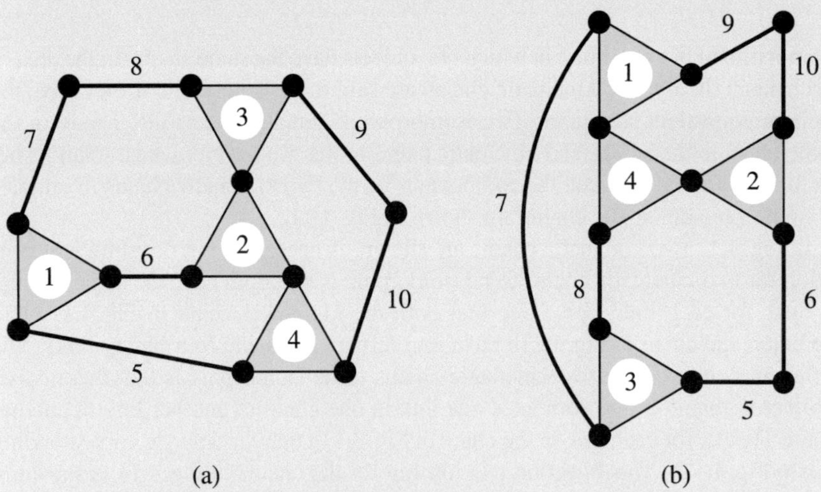

Fig. 4.12 Isomorphism between kinematic chains

Solution:

The number of links in both chains is $n = 10$, and the number of kinematic pairs in both is $j = 12$. Thus, the chains have mobility 3. Since degree of control, connectivity and redundancy are internal properties of one link relative to the other, the highest possible degree of control in these kinematic chains is 3. Thus, no element of the degree of control matrix should have a value greater than 3.

Figure chain 4.12**a**

Degree of control matrix

$$K = \begin{bmatrix} 0&2&3&2&1&1&1&2&3&3 \\ 2&0&1&1&2&1&3&2&2&2 \\ 3&1&0&2&3&2&2&1&1&2 \\ 2&1&2&0&1&2&3&3&2&1 \\ 1&2&3&1&0&2&2&3&3&2 \\ 1&1&2&2&2&0&2&3&3&3 \\ 1&3&2&3&2&2&0&1&3&3 \\ 2&2&1&3&3&3&1&0&2&3 \\ 3&2&1&2&3&3&3&2&0&1 \\ 3&2&2&1&2&3&3&3&1&0 \end{bmatrix} \tag{4.12}$$

Connectivity matrix

$$C = \begin{bmatrix} 0&2&3&2&1&1&1&2&3&3 \\ 2&0&1&1&2&1&3&2&2&2 \\ 3&1&0&2&3&2&2&1&1&2 \\ 2&1&2&0&1&2&3&3&2&1 \\ 1&2&3&1&0&2&2&3&3&2 \\ 1&1&2&2&2&0&2&3&3&3 \\ 1&3&2&3&2&2&0&1&3&3 \\ 2&2&1&3&3&3&1&0&2&3 \\ 3&2&1&2&3&3&3&2&0&1 \\ 3&2&2&1&2&3&3&3&1&0 \end{bmatrix} \tag{4.13}$$

Redundancy matrix

$$R = 0_{10\times 10} \tag{4.14}$$

Figure chain 4.12**b** Degree of control matrix

$$K = \begin{bmatrix} 0&2&2&1&3&3&1&2&1&2 \\ 2&0&3&1&2&1&3&2&2&1 \\ 2&3&0&2&1&2&1&1&3&3 \\ 1&1&2&0&3&2&2&1&2&2 \\ 3&2&1&3&0&1&2&2&3&3 \\ 3&1&2&2&1&0&3&3&3&2 \\ 1&3&1&2&2&3&0&2&2&3 \\ 2&2&1&1&2&3&2&0&3&3 \\ 1&2&3&2&3&3&2&3&0&1 \\ 2&1&3&2&3&2&3&3&1&0 \end{bmatrix} \tag{4.15}$$

Connectivity matrix

$$C = \begin{bmatrix} 0&2&2&1&3&3&1&2&1&2 \\ 2&0&3&1&2&1&3&2&2&1 \\ 2&3&0&2&1&2&1&1&3&3 \\ 1&1&2&0&3&2&2&1&2&2 \\ 3&2&1&3&0&1&2&2&3&3 \\ 3&1&2&2&1&0&3&3&3&2 \\ 1&3&1&2&2&3&0&2&2&3 \\ 2&2&1&1&2&3&2&0&3&3 \\ 1&2&3&2&3&3&2&3&0&1 \\ 2&1&3&2&3&2&3&3&1&0 \end{bmatrix} \tag{4.16}$$

Redundancy matrix

$$R = 0_{10\times 10} \tag{4.17}$$

Comparison

The kinematic chains graphs are shown in Fig. 4.13. It can be seen that the kinematic chain isomorphism is repeated in the graphs as well.

As expected, the degree of control and connectivity matrices are not the same; since the links have swapped numbering, the rows and columns of such matrices will also be swapped. However, these matrices do have some similarities. First row of the K matrix in Fig. 4.12a has one null element, three elements 1, three elements 2, and three elements 3. This row corresponds to the third row of the K matrix in Fig. 4.12b. Such a relationship can be constructed for all rows of the degree of control and connectivity matrices.

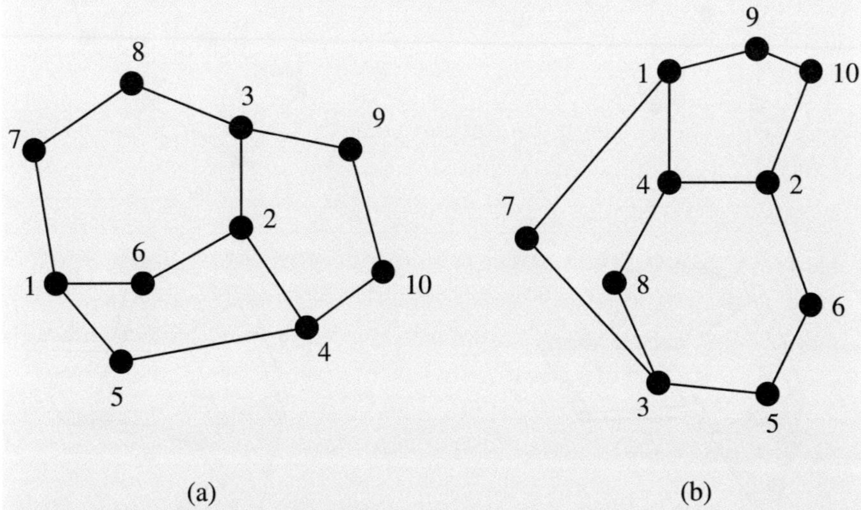

Fig. 4.13 Isomorphism between graphs

For more information regarding isomorphism, see [5, 6].

Fig. 4.14 Planar kinematic chain

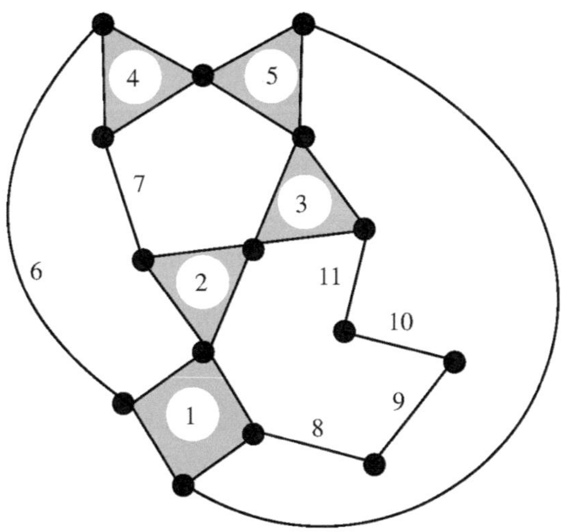

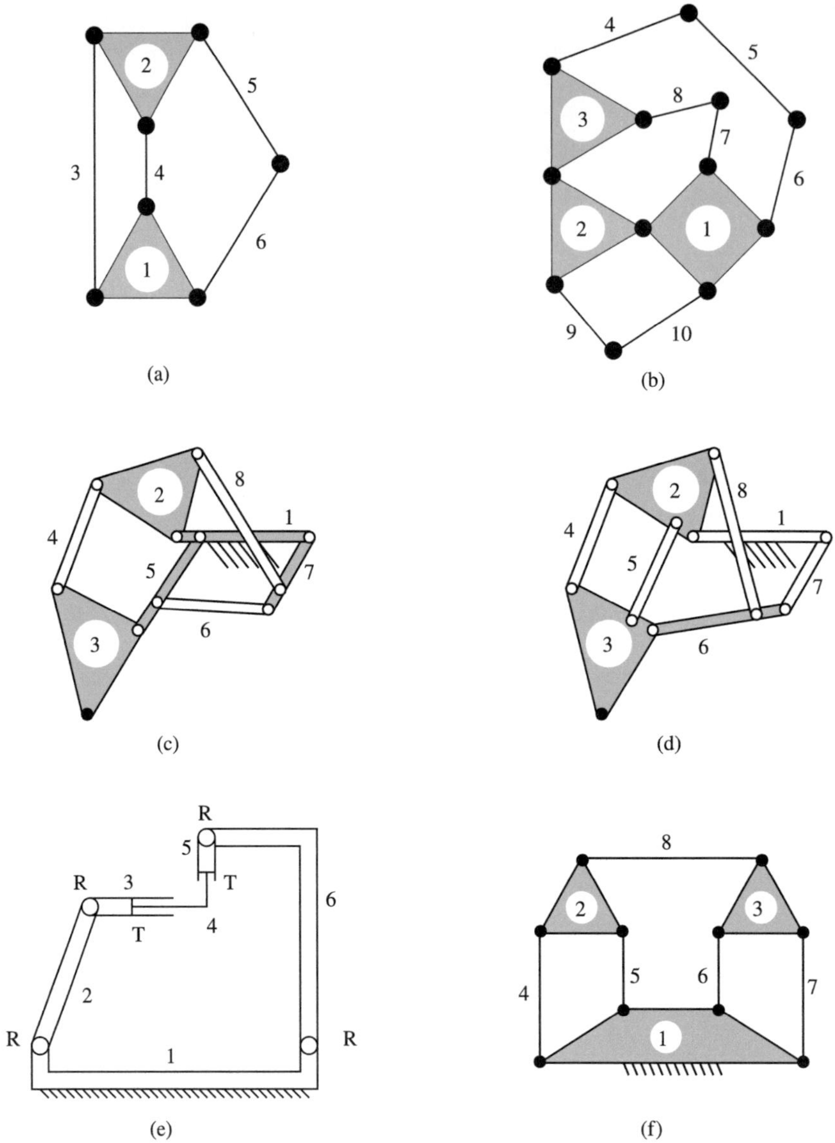

Fig. 4.15 Kinematic chains and planar mechanisms

Problems

4.1 Determine the redundancy matrix for Example 4.

4.2 For a planar mechanism with mobility M, what is the largest possible value for variety?

4.3 For the kinematic chain in Fig. 4.14:

(a) Determine the kinematic chain mobility.
(b) If the kinematic chain is planar and considering the mobility calculated in the previous item, what can be stated about the degree of control and connectivity matrices?
(c) Considering the previous item, what can be stated about the redundancy matrix?
(d) Determine the connectivity, degree of control, and redundancy matrices.
(e) Determine the variety.

4.4 For the kinematic chains and planar mechanisms in Fig. 4.15, determine the graph; the degree of control, connectivity and redundancy matrices; and the variety.

References

1. Carboni, A.P.: Análise conceitual de estruturas cinemáticas planas e espaciais. Master's dissertation. Federal University of Santa Catarina, Florianópolis (2008)
2. Martins, D., Carboni, A.P.: Variety and connectivity in kinematic chains. Mech. Mach. Theory **43**(10), 1236–1252 (2008)
3. Carboni, A.P., Martins, D.: Redundancy and connectivity in kinematic chains. In: Proceedings: 19th International Congress of Mechanical Engineering. COBEM, Brasília (2007)
4. Belfiore, N.P., Benedetto, A.D.: Connectivity and redundancy in spatial robots. Int. J. Rob. Res. **19**(12), 1245–1261 (2000)
5. Mruthyunjaya, T.S.: Kinematic structure of mechanisms revisited. Mech. Mach. Theory **38**(4), 279–320 (2003)
6. Simoni, R.: Síntese estrutural de cadeias cinemáticas e mecanismos. Master's dissertation. Federal University of Santa Catarina, Florianópolis (2008)

Chapter 5
Kinematic Analysis

Abstract This chapter provides a comprehensive exploration of kinematic analysis, covering position, velocity, and acceleration analyses for mechanisms. The position analysis section introduces geometric methods and vector loop equations to determine links and points locations for given input values. Velocity and acceleration analyses follow, incorporating relative velocity equations and differentiation techniques to evaluate dynamic behaviors. Special attention is given to Jacobian matrices for analyzing motion at specific points of interest, which is critical in robotic and control applications. Examples, including planar mechanisms with revolute and prismatic pairs, help illustrate the analytical process. The chapter also touches on more advanced configurations like quick-return mechanisms.

Keywords Kinematic analysis · Position analysis · Velocity analysis · Acceleration analysis · Point of interest

This chapter will cover the kinematic analysis of mechanisms. Initially, basic concepts are covered, then the position analysis will be looked at, followed by the velocity analysis, and finally the acceleration analysis. The Jacobian and the point of interest analysis will also be covered. More examples and details regarding kinematic analysis can be seen at [1–3].

5.1 Position Analysis

The **position analysis** determines the position and orientation of each component of a mechanism, given values for its degrees of freedom. For example, given the input angle value of a four-bar mechanism, studied in Chap. 3, position analysis provides the location and orientation of each link of the four-bar mechanism.

For didactic purposes, the position analysis for the slider-crank mechanism is presented first. More examples are presented in Sect. 5.6.

Connecting rod (*connecting rod*) is a component that contains a revolute pair whose axis is subject to translation. An example is shown in Fig. 5.1.

D. Martins and E. H. Murai, *Mechanisms–Synthesis and Analysis with Applications in Robotics*, Mechanisms and Machine Science 183,
https://doi.org/10.1007/978-3-031-95187-9_5

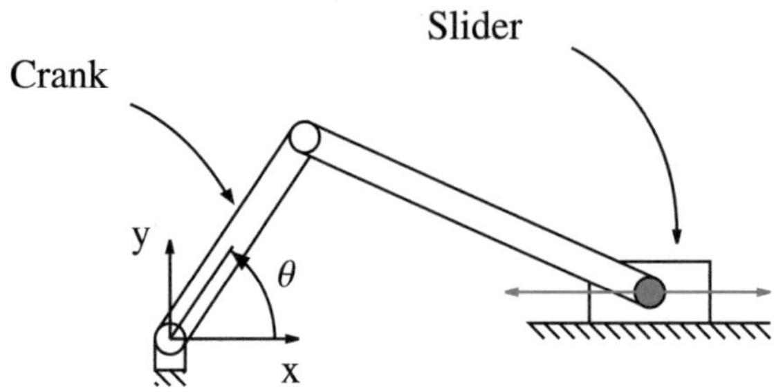

Fig. 5.1 Slider-crank mechanism

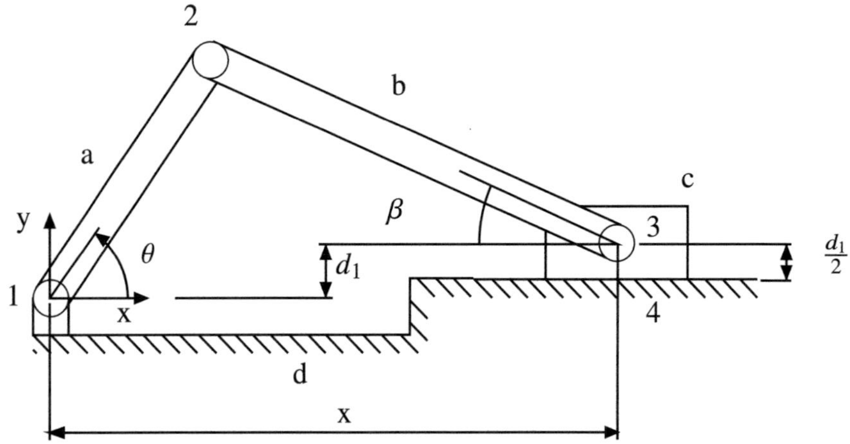

Fig. 5.2 Slider-crank mechanism

A slider-crank mechanism is shown in Fig. 5.1. This mechanism has three links, the crank, the connecting rod and the slider. The slider link is also called the piston.

In Fig. 5.2, one can determine the position of kinematic pair 2 relative to pair 1. This position is given by

$$\begin{aligned} x_{21} &: \quad a\cos(\theta) \\ y_{21} &: \quad a\sin(\theta). \end{aligned} \tag{5.1}$$

The position of pair 3 relative to pair 2 is given by

$$\begin{aligned} x_{32} &: \quad b\cos(\beta) \\ y_{32} &: \quad -b\sin(\beta). \end{aligned} \tag{5.2}$$

The position of pair 4 relative to pair 3 is given by

$$\begin{aligned} x_{43} &: \quad 0 \\ y_{43} &: \quad -\frac{d_1}{2}. \end{aligned} \tag{5.3}$$

The position of pair 1 relative to pair 4 is given by

$$\begin{aligned} x_{14} &: \quad -x \\ y_{14} &: \quad -\frac{d_1}{2}. \end{aligned} \tag{5.4}$$

To determine the position of pair 3 relative to pair 1, one simply adds the position of 3 relative to 2 and of 2 relative to 1. So, adding Eqs. 5.1 and 5.2, we have

$$\begin{aligned} x_{31} &: \quad a\cos(\theta) + b\cos(\beta) \\ y_{31} &: \quad a\sin(\theta) - b\sin(\beta). \end{aligned} \tag{5.5}$$

Similarly, adding Eqs. 5.1, 5.2, 5.3 and 5.4, gives

$$\begin{aligned} x_{11} &: \quad a\cos(\theta) + b\cos(\beta) - x \\ y_{11} &: \quad a\sin(\theta) - b\sin(\beta) - \frac{d_1}{2} - \frac{d_1}{2}. \end{aligned} \tag{5.6}$$

Equation 5.6 determine the position of kinematic pair 1 relative to pair 1 itself. The distance of a point relative to itself is 0, so one has

$$\begin{aligned} x &: \quad a\cos(\theta) + b\cos(\beta) - x = 0 \\ y &: \quad a\sin(\theta) - b\sin(\beta) - d_1 = 0. \end{aligned} \tag{5.7}$$

Equation 5.7 are often called *equation of the position in* x and *equation of the position in* y.

In the Eq. 5.7, the variables a, b, d_1 and θ are known, and one has to determine the variables β and x. Since there are two equations (x and y) and two unknowns (β and x), it is possible to solve this system in an unique way. Rearranging the terms of Eq. 5.7, one has

$$\begin{aligned} x &= a\cos(\theta) + b\cos(\beta) \\ \beta &= \arcsin\left(\frac{a\sin(\theta) - d_1}{b}\right). \end{aligned} \tag{5.8}$$

Thus, the values of x and β for any input angle θ are determined. Using $a = 2\, l.u.$, $b = 4\ l.u.$ and $d_1 = 0.5\ l.u.$, one obtains the graphs shown in Figs. 5.3 and 5.4. The script used to determine the graphs of the slider-crank kinematics is shown in Appendix B.

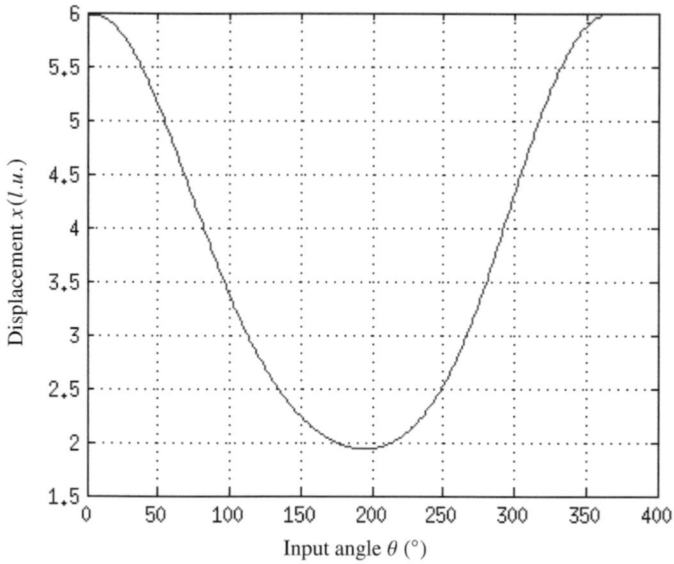

Fig. 5.3 Piston displacement graph $x(\theta)$

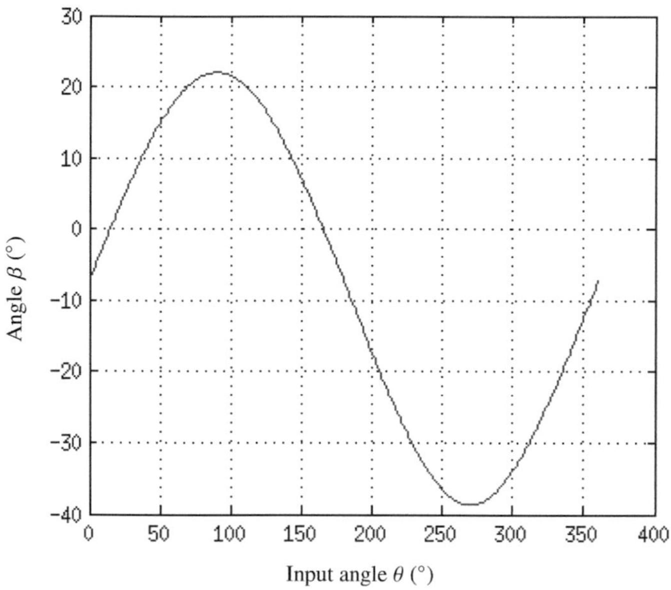

Fig. 5.4 Connecting rod angle graph $\beta(\theta)$

5.2 Velocity Analysis

The **velocity analysis** determines the linear and angular velocity of each component of a mechanism, given values for the mechanism degrees of freedom and their rate of change in time.

The Eq. 5.7 are written again as follows

$$\begin{aligned} x: \quad & a\cos(\theta) + b\cos(\beta) - x = 0 \\ y: \quad & a\sin(\theta) - b\sin(\beta) - d_1 = 0. \end{aligned}$$

By deriving these equations with respect to time, we get the velocity equations

$$\begin{aligned} -a\sin(\theta)\dot{\theta} - b\sin(\beta)\dot{\beta} - \dot{x} &= 0 \\ a\cos(\theta)\dot{\theta} - b\cos(\beta)\dot{\beta} &= 0. \end{aligned} \tag{5.9}$$

Isolating the desired variables, $\dot{x}$ and $\dot{\beta}$, yields

$$\begin{aligned} \dot{x} &= -a\sin(\theta)\dot{\theta} - b\sin(\beta)\dot{\beta} \\ \dot{\beta} &= \frac{a\cos(\theta)\dot{\theta}}{b\cos(\beta)}. \end{aligned} \tag{5.10}$$

In order to solve Eq. 5.10, the values of θ, $\dot{\theta}$ and β are required, as well as the mechanism dimensions. Thus, it is necessary to solve the position kinematics before solving the velocity kinematics. For the same slider-crank used in the Sect. 5.1, making $\dot{theta} = 1\ rad/s$, one obtains the graphs shown in Figs. 5.5 and 5.6.

Another convenient way to write the Eq. 5.9 is using matrices

$$\begin{bmatrix} -b\sin(\beta) & -1 \\ -b\cos(\beta) & 0 \end{bmatrix} \begin{Bmatrix} \dot{\beta} \\ \dot{x} \end{Bmatrix} = \begin{Bmatrix} a\sin(\theta)\dot{\theta} \\ -a\cos(\theta)\dot{\theta} \end{Bmatrix}. \tag{5.11}$$

In Eq. 5.11, the matrix establishes a relationship between the input velocity ($\dot{\theta}$ on the right side of the equation) and the output velocity ($\dot{\beta}$ and $\dot{x}$ on the left side of the equation). This matrix that relates input parameters to output parameters is called **Jacobian matrix**.[1]

Jacobian is a composite matrix of the partial derivatives of a vector function. Considering the Eq. 5.9 and taking their partial derivatives, one can determine the Jacobian matrix.

$$J = \begin{bmatrix} \frac{\partial F_1}{\partial x_1} & \frac{\partial F_1}{\partial x_2} \\ \frac{\partial F_2}{\partial x_1} & \frac{\partial F_2}{\partial x_2} \end{bmatrix} = \begin{bmatrix} \frac{\partial F_1}{\partial \dot{\beta}} & \frac{\partial F_1}{\partial \dot{x}} \\ \frac{\partial F_2}{\partial \dot{\beta}} & \frac{\partial F_2}{\partial \dot{x}} \end{bmatrix} = \begin{bmatrix} -b\sin(\beta) & -1 \\ -b\cos(\beta) & 0 \end{bmatrix}. \tag{5.12}$$

[1] Named after the mathematician Carl Gustav Jakob Jacobi (1804–1851).

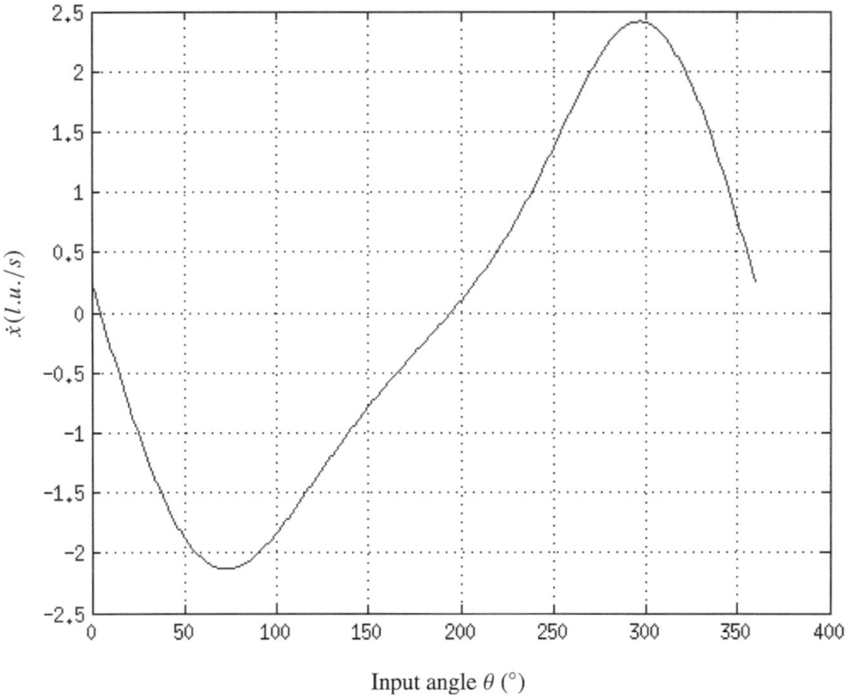

Fig. 5.5 Piston velocity graph $\dot{x}$

Mathematically, one can solve Eq. 5.11 by multiplying both sides by the inverse of the Jacobian. However, such an inverse will not always exist. For example, when the determinant of the Jacobian matrix is zero, there will be no inverse matrix, and thus it is not possible to solve the system of Eq. 5.11.

Physically, when the determinant of the Jacobian is zero, the mechanism is at a singularity point. For example, if $\beta = 90°$, the second row of the matrix in Eq. 5.11 cancels out, so the determinant is zero. In this case, the mechanism would be in the position shown in Fig. 5.7. Note that not every slider-crank will reach the position at which $\beta = 90°$, and the mechanism dimensions determine the limits of β.

Analyzing Fig. 5.7, one can see that by reducing θ it is not possible to determine whether the link c will go left or right. Dynamically, one could argue that the link would leave the singularity position tending to follow its motion by inertia. However, one must remember that in the scope of kinematic analysis, forces and moments are not considered. Furthermore, low velocities and/or the presence of friction forces can overcome inertial forces.

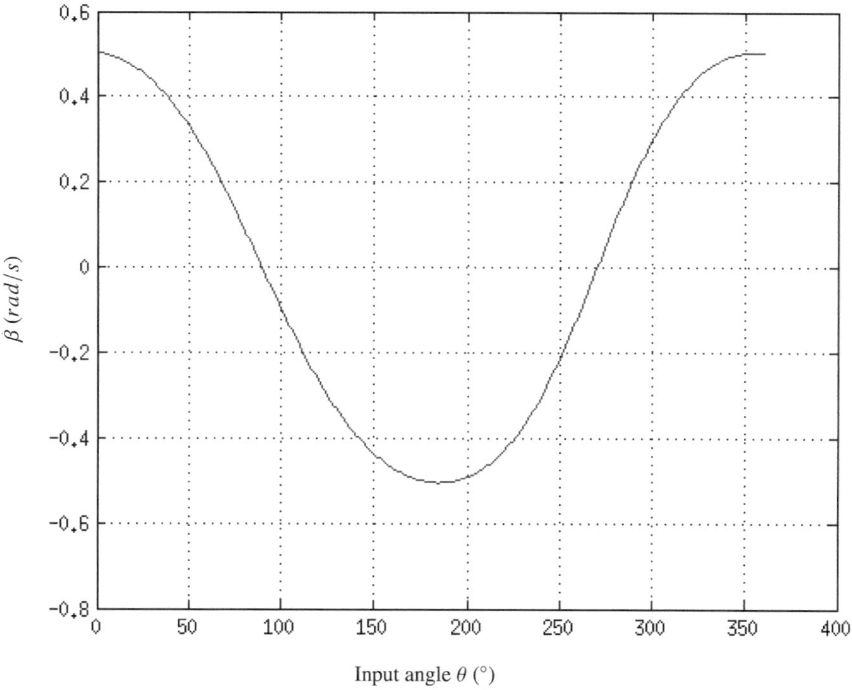

Fig. 5.6 Graph of angular velocity of the connecting rod $\dot{\beta}$

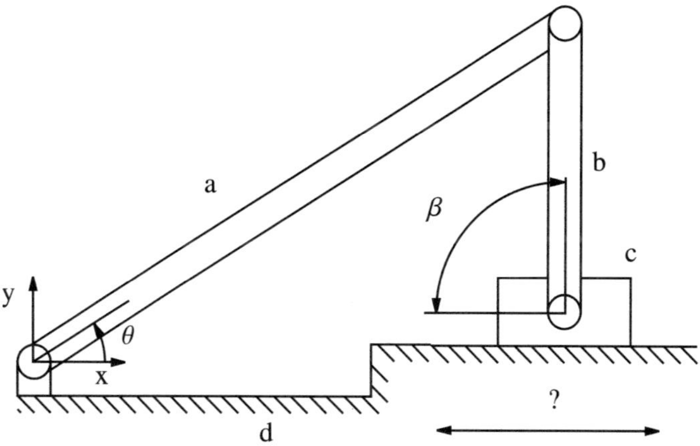

Fig. 5.7 Slider-crank in a singular position

5.3 Acceleration Analysis

The **acceleration analysis** defines the linear and angular acceleration of each component of a mechanism, given the values for the mechanism degrees of freedom, their velocities, and accelerations.

Equation 5.9 are shown again as follows

$$\begin{aligned} -a\sin(\theta)\dot{\theta} - b\sin(\beta)\dot{\beta} - \dot{x} &= 0 \\ a\cos(\theta)\dot{\theta} - b\cos(\beta)\dot{\beta} &= 0. \end{aligned}$$

By deriving with respect to time such equations, one obtains the equations for the acceleration

$$\begin{aligned} -a\cos(\theta)\dot{\theta}^2 - a\sin(\theta)\ddot{\theta} - b\cos(\beta)\dot{\beta}^2 - b\sin(\beta)\ddot{\beta} - \ddot{x} &= 0 \\ -a\sin(\theta)\dot{\theta}^2 + a\cos(\theta)\ddot{\theta} + b\sin(\beta)\dot{\beta}^2 - b\cos(\beta)\ddot{\beta} &= 0. \end{aligned} \quad (5.13)$$

Isolating the desired variables

$$\begin{aligned} \ddot{x} &= -a\cos(\theta)\dot{\theta}^2 - a\sin(\theta)\ddot{\theta} - b\cos(\beta)\dot{\beta}^2 - b\sin(\beta)\ddot{\beta} \\ \ddot{\beta} &= \frac{-a\sin(\theta)\dot{\theta}^2 + a\cos(\theta)\ddot{\theta} + b\sin(\beta)\dot{\beta}^2}{b\cos(\beta)}. \end{aligned} \quad (5.14)$$

To solve Eq. 5.14, the values of θ, $\dot{\theta}$, $\ddot{\theta}$, β and $\dot{\beta}$ are required, as well as the mechanism dimensions. Thus, it is necessary to solve the velocity kinematics before solving the acceleration kinematics. For the same slider-crank used in Sect. 5.1, by making $\dot{\theta} = 1\ rad/s$ and $\ddot{\theta} = 0\ rad/s^2$, the graphs shown in Figs. 5.8 and 5.9 are obtained.

Another convenient way to write the Eq. 5.13 is using matrices

$$\begin{bmatrix} -b\sin(\beta) & -1 \\ -b\cos(\beta) & 0 \end{bmatrix} \begin{Bmatrix} \ddot{\beta} \\ \ddot{x} \end{Bmatrix} = \begin{Bmatrix} a\cos(\theta)\dot{\theta}^2 + a\sin(\theta)\ddot{\theta} + b\cos(\beta)\dot{\beta}^2 \\ a\sin(\theta)\dot{\theta}^2 - a\cos(\theta)\ddot{\theta} - b\sin(\beta)\dot{\beta}^2 \end{Bmatrix}. \quad (5.15)$$

Again, it can be seen that in Eq. 5.15 the Jacobian matrix appears. The recursion of the Jacobian matrix serves to aid in checking the math, since the matrix must be the same in both Eq. 5.15 and Eq. 5.11.

5.4 Displacement, Path and Trajectory

This section deals with correlated mechanism terms often used in robotics.

Displacement is the change in position of a point or material body. For example, if a point P translates from position 1 to position 2, the displacement is given by

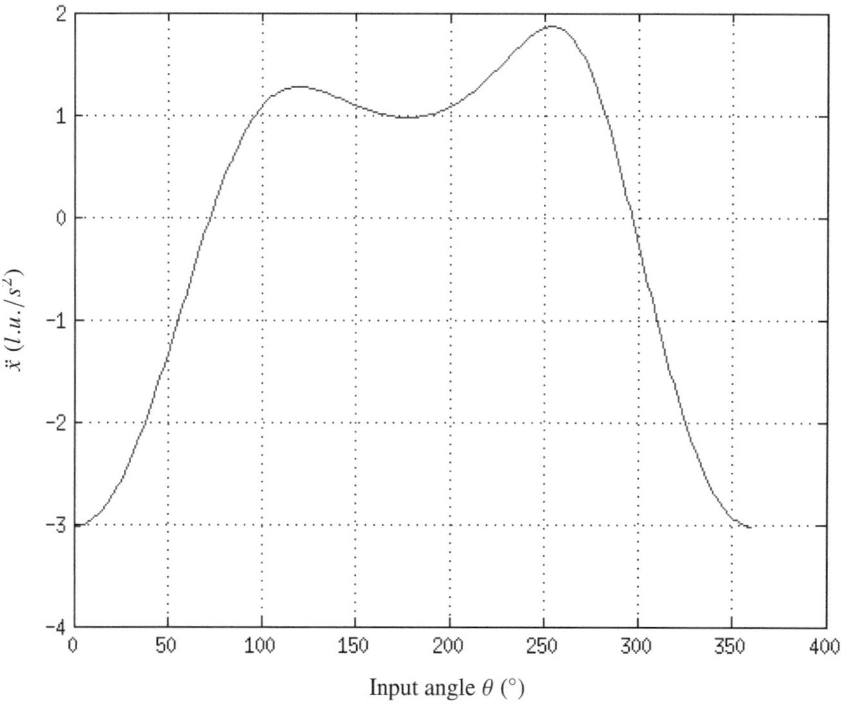

Fig. 5.8 Piston acceleration graph $\ddot{x}$

$\mathbf{r}_{12} = \mathbf{r}_2 - \mathbf{r}_1$, as shown in Fig. 5.10a. If a link is rotating from a start position θ_1 to an end position θ_2, its displacement is given by $\theta_{12} = \theta_2 - \theta_1$, as can be seen in Fig. 5.10b.

In some mechanism and robotics problems, a displacement specification is enough to satisfactorily model the task to be done, such as *pick-and-place* type operations. In a pick-and-place operation, it is desired to take the object from one place and deposit it somewhere else. For example, taking the object off a conveyor belt and depositing it on a pallet. Thus, to a robot or a mechanism do such a task, it must pass through the two points, start and end, no matter which *path* is traveled between these two points.

Path is the displacement measured over time. The difference between displacement and path is that in displacement only the start and end positions matter, whereas path takes into account how the final position was reached, i.e. , it also considers intermediate positions. In the example of a point moving in space, there are infinitely many possible paths connecting the point 1 to the point 2, as can be seen in Fig. 5.11a. For the rotating link example, the link can reach the θ_2 position by either a clockwise rotation or a counterclockwise rotation, as shown in Fig. 5.11b.

A *pick-and-place* operation can also be modeled via a path, having the advantage of being able to impose the shortest path in order to optimize the process. Another

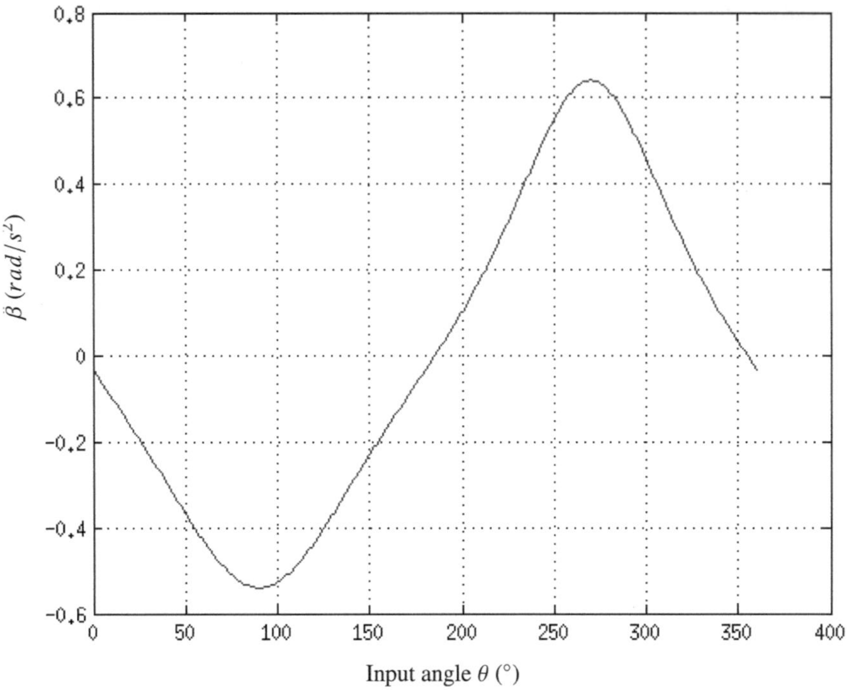

Fig. 5.9 Graph of angular acceleration of the connecting rod $\ddot{\beta}$

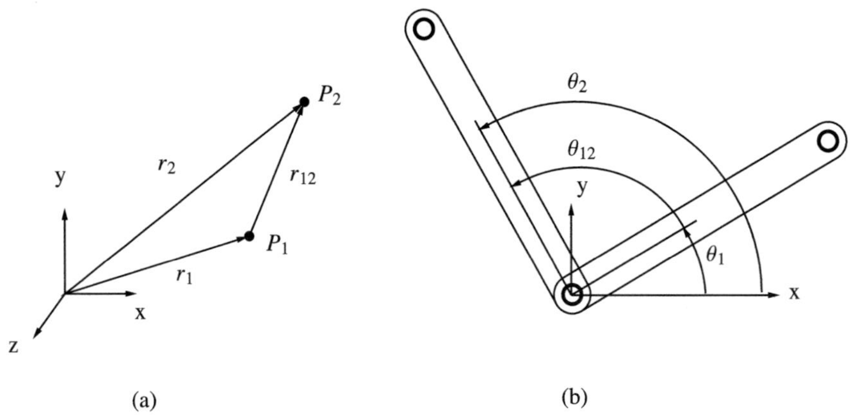

Fig. 5.10 Different displacements. **a** Displacement of a point in space. **b** Displacement of a link in the plane

advantage of specifying *pick-and-place* operations using path instead of displacement is to avoid collisions. Thus, you can plan the path in such a way that it deflects any object that is within the robot's workspace.

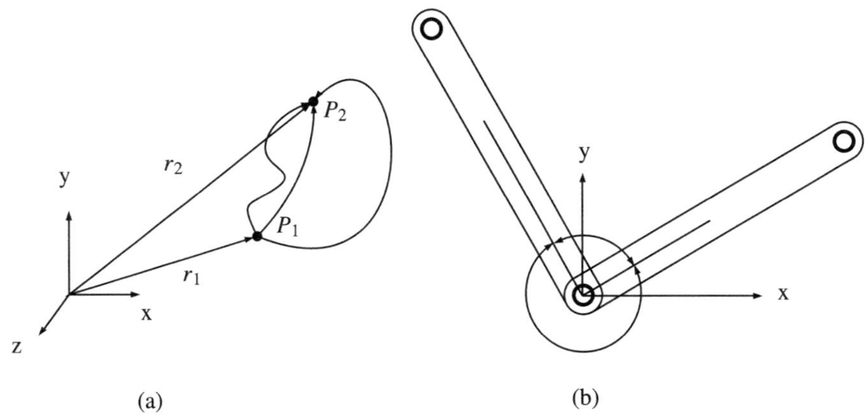

Fig. 5.11 Different paths. **a** Possible paths for moving a point in space. **b** Possible paths for rotating a link in the plane

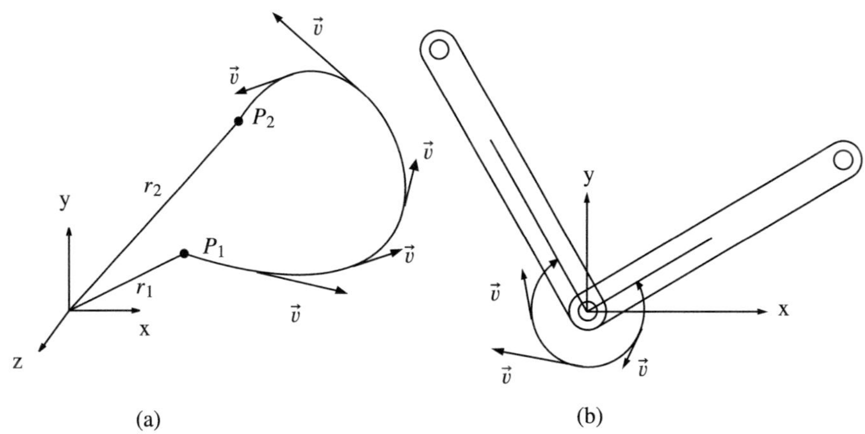

Fig. 5.12 Different trajectories. **a** Trajectory for the path of a point in space. **b** Trajectory for the path of a link in the plane

Trajectory is a path with prescribed velocities. In addition to the concept of a path, which considers intermediate positions, the trajectory also considers the velocities at intermediate positions. In Fig. 5.12 a possible trajectory for the path example is shown.

Welding is an example of a task that must be specified using trajectory. In the welding process, the welding tool must travel along the path to be welded with a specified speed. If the speed is slower than desired, the material in the vicinity of the weld will overheat, if the speed is faster than desired, the material will not heat up enough to make a good weld. Another example where speed must be considered are conveyor operations, such as *palletizing*, use of automated forklifts, and use of an AGV (*Automatic Guided Vehicle*). Such operations should have smooth speed changes in order to avoid damage to the transported material and even its fall.

5.5 Point of Interest Analysis

As stated in Chap. 1, one often needs to determine the kinematics of a specific point belonging to one of the links of the mechanism.

For example, a convertible roof mechanism is shown in Fig. 5.13a. For a convertible roof, it is desired to know the extreme points kinematics (circled in Fig. 5.13a). These points present locks that engage the windshiled frame, thus, determining their path is necessary when designing the roof.

Another example of a point of interest is the slider-crank mechanism used to push hay into a baler. Such mechanism is shown in Fig. 5.13b. In this mechanism, the extensions of the connecting rod have the function of pushing the hay toward the baler.

Consider the connecting rod with an added point P belonging to the connecting rod, as shown in Fig. 5.14.

The x_1y_1 coordinate system is a moving referential to aid in the formulation development. Like point P, the x_1y_1 coordinate system belongs to the connecting rod and moves along with it. Thus, the position of point P relative to the system x_1y_1, $P(x_{1p}, y_{1p})$ is time invariant (a feature that will facilitate the derivation with respect to time).

Analyzing Fig. 5.15, one can determine the position of the point P relative to the fixed referential, x_p, as being

$$x_p = a\cos(\theta) + y_{1p}\sin(\beta) + x_{1p}\cos(\beta). \tag{5.16}$$

Analyzing Fig. 5.16, one can determine the position of the point P with respect to the fixed referential, y_p, as being

$$y_p = a\sin(\theta) + y_{1p}\cos(\beta) - x_{1p}\sin(\beta). \tag{5.17}$$

Equations 5.16 and 5.17 can be rewritten using matrices

$$\begin{Bmatrix} x_p \\ y_p \end{Bmatrix} = \begin{Bmatrix} a\cos(\theta) \\ a\sin(\theta) \end{Bmatrix} + \begin{bmatrix} \cos(\beta) & \sin(\beta) \\ -\sin(\beta) & \cos(\beta) \end{bmatrix} \begin{Bmatrix} x_{1p} \\ y_{1p} \end{Bmatrix}. \tag{5.18}$$

The first vector in Eq. 5.18 refers to the **position** of the moving referential (system x_1y_1) relative to the fixed referential (system xy). The matrix in Eq. 5.18 refers to the **orientation** of the moving referential (system x_1y_1) relative to the fixed referential (system xy).

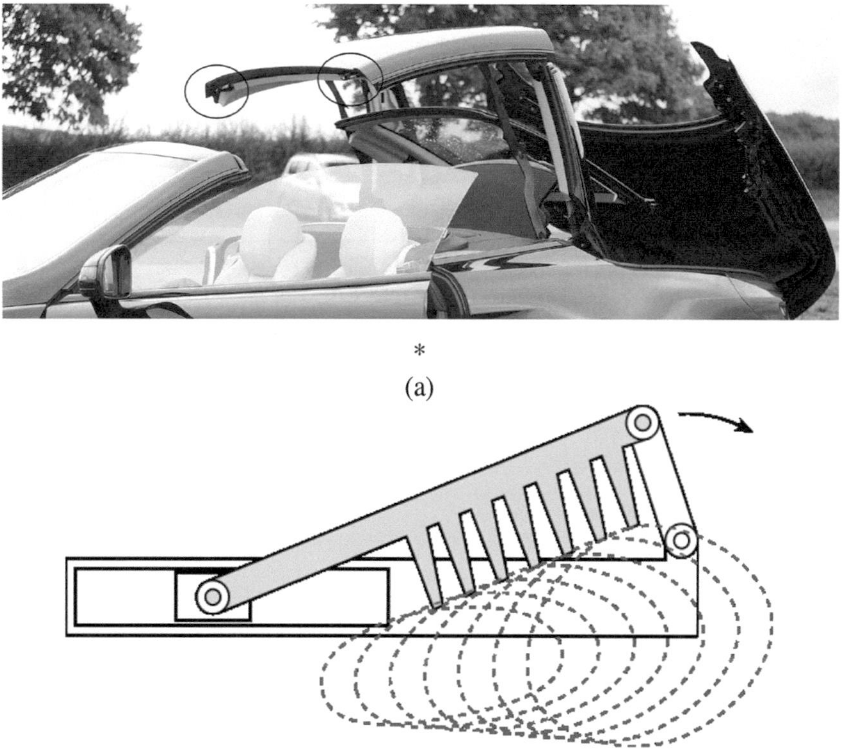

(a)

(b)

Fig. 5.13 Examples of point of interest analysis. **a** Convertible roof mechanism. **b** Baler mechanism

To determine the position of P in relation to the fixed referential we need to know the position of the point P in relation to the moving referential (x_{1p} and y_{1p}) and the angle β. Thus, one must solve the position kinematics before analyzing the point of interest.

To determine the velocity of the point of interest, simply derive with respect to time the Eqs. 5.16 and 5.17:

$$\begin{aligned} \dot{x_p} &= -a\sin(\theta)\dot{\theta} + y_{1p}\cos(\beta)\dot{\beta} - x_{1p}\sin(\beta)\dot{\beta} \\ \dot{y_p} &= \ \ a\cos(\theta)\dot{\theta} - y_{1p}\sin(\beta)\dot{\beta} - x_{1p}\cos(\beta)\dot{\beta}. \end{aligned} \tag{5.19}$$

To determine the acceleration of the point of interest, simply derive the Eq. 5.19 with respect to time:

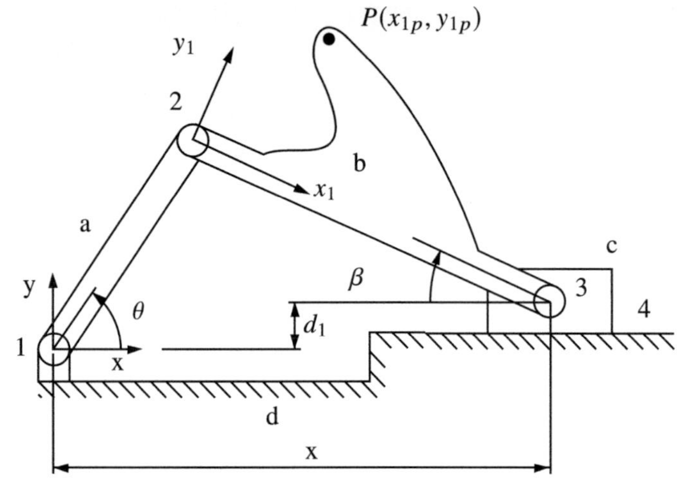

Fig. 5.14 Slider-crank with point of interest

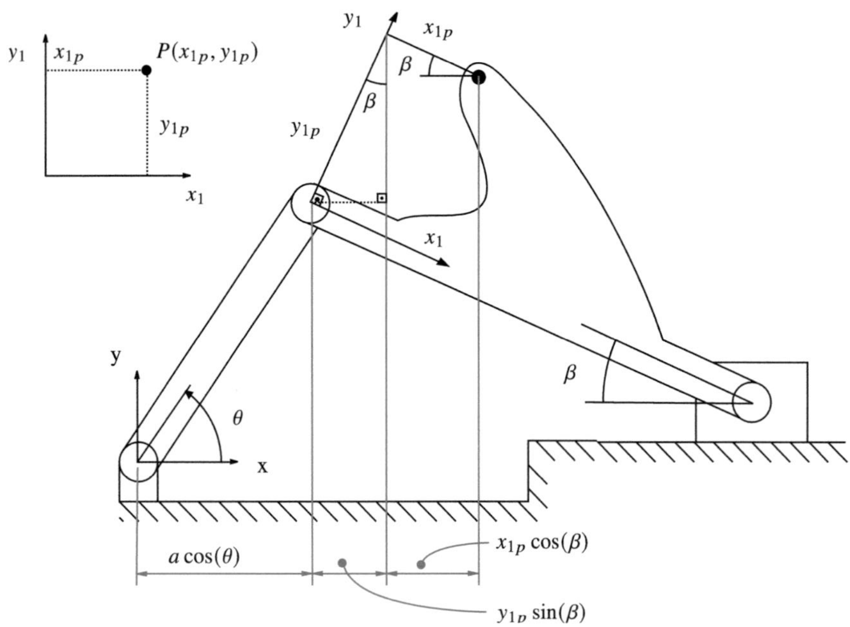

Fig. 5.15 x-coordinate of P with respect to the fixed referential

$$
\begin{aligned}
\ddot{x_p} &= -a\cos(\theta)\dot{\theta}^2 - a\sin(\theta)\ddot{\theta} - y_{1p}\sin(\beta)\dot{\beta}^2 + y_{1p}\cos(\beta)\ddot{\beta} \\
&\quad -x_{1p}\cos(\beta)\dot{\beta}^2 - x_{1p}\sin(\beta)\ddot{\beta} \\
\ddot{y_p} &= -a\sin(\theta)\dot{\theta}^2 + a\cos(\theta)\ddot{\theta} - y_{1p}\cos(\beta)\dot{\beta}^2 - y_{1p}\sin(\beta)\ddot{\beta} \\
&\quad +x_{1p}\sin(\beta)\dot{\beta}^2 - x_{1p}\cos(\beta)\ddot{\beta}.
\end{aligned}
\tag{5.20}
$$

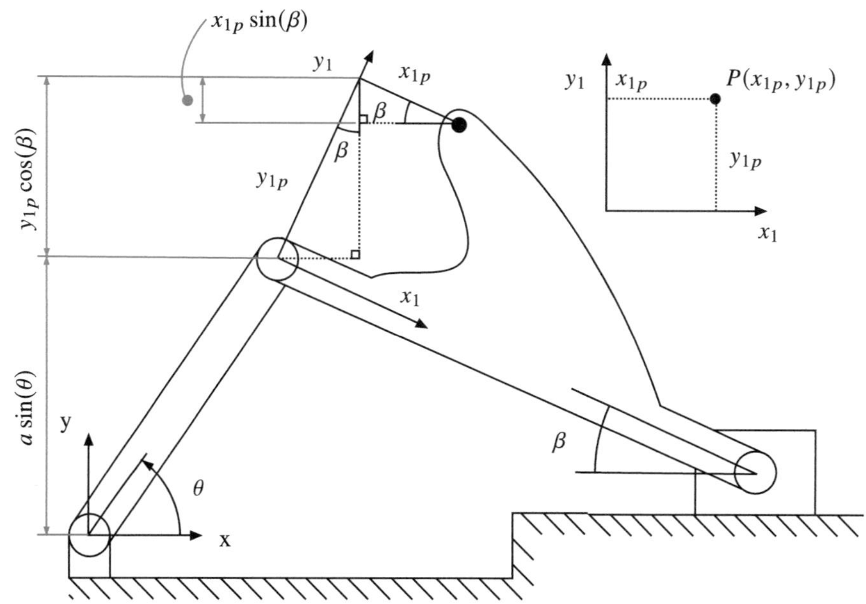

Fig. 5.16 y-coordinate of P with respect to the fixed referential

To determine the velocity of the point of interest, the velocity kinematics for the slider-crank must first be solved. Similarly, to determine the acceleration of the point of interest, one needs to solve the acceleration kinematics for the slider-crank.

5.6 Examples

In this section, three examples are presented. The goal is for the reader to have a better understanding of how to solve the kinematics in various cases, such as when there is more than one loop or more than one mobility.

5.6.1 Four-Bar Mechanism

Consider the four-bar mechanism shown in Fig. 5.17.

Position analysis
Analyzing the loop in x and in y, one has the position equations

$$\begin{aligned} x: \quad & a\cos(\theta) + b\cos(\alpha) + c\cos(\beta) - d = 0 \\ y: \quad & a\sin(\theta) + b\sin(\alpha) + c\sin(\beta) \quad\;\; = 0 \end{aligned} \tag{5.21}$$

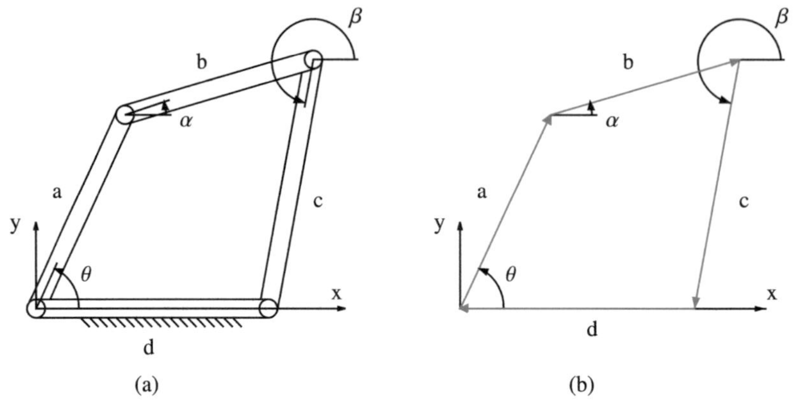

Fig. 5.17 Four-bar mechanism. **a** Functional representation. **b** Vector representation

In Eq. 5.21, α and β are the desired variables. The Eq. 5.21 have no analytical solution and must be solved numerically. To solve them, one can adopt a value for α and calculate the value of β in the equations for x and y. If both equations result in the same value for β, then the chosen value for α is a solution. Otherwise, one must change the value of α until both β have the same value (or a close value, according to a tolerance). This is an iterative process, and some computer program must be used to solve the system of equations. One can also use any other numerical method, such as Newton-Raphson. The reader should remember that the four-bar mechanism has two working modes, so whatever solving method is chosen, one should draft the position of the mechanism in the desired working mode. When entering a starting point to solve numerically, one uses the draft to provide a starting point closer to the desired solution and thus prevent the numerical method from converging to the undesired working mode.

Velocity analysis

Deriving Eq. 5.21 with respect to time, one has

$$\begin{aligned} x: \quad & -a\sin(\theta)\dot{\theta} - b\sin(\alpha)\dot{\alpha} - c\sin(\beta)\dot{\beta} = 0 \\ y: \quad & a\cos(\theta)\dot{\theta} + b\cos(\alpha)\dot{\alpha} + c\cos(\beta)\dot{\beta} = 0. \end{aligned} \tag{5.22}$$

Rewriting the Eq. 5.22, using matrix

$$\begin{bmatrix} b\sin(\alpha) & c\sin(\beta) \\ b\cos(\alpha) & c\cos(\beta) \end{bmatrix} \begin{Bmatrix} \dot{\alpha} \\ \dot{\beta} \end{Bmatrix} = \begin{Bmatrix} -a\sin(\theta)\dot{\theta} \\ -a\cos(\theta)\dot{\theta} \end{Bmatrix}. \tag{5.23}$$

Note that the matrix of Eq. 5.23 is the Jacobian matrix.

Acceleration analysis

Deriving again with respect to time we get the accelerations.

Fig. 5.18 Coupler point

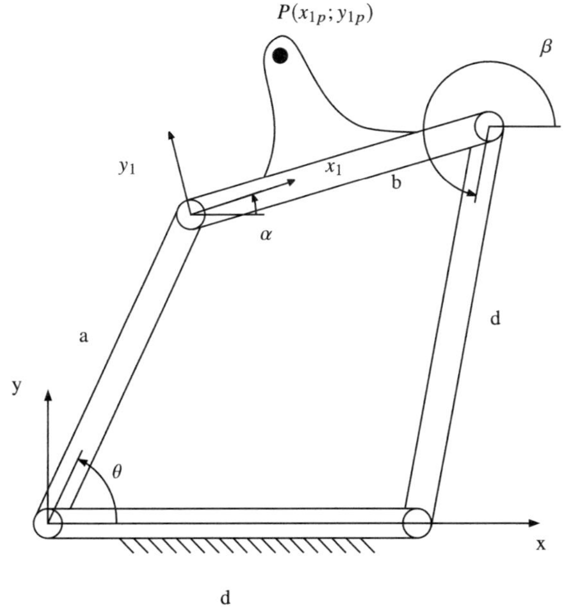

$$
\begin{aligned}
x: \quad & -a\cos(\theta)\dot{\theta}^2 - a\sin(\theta)\ddot{\theta} - b\cos(\alpha)\dot{\alpha}^2 - b\sin(\alpha)\ddot{\alpha} \\
& \quad -c\cos(\beta)\dot{\beta}^2 - c\sin(\beta)\ddot{\beta} = 0 \\
y: \quad & -a\sin(\theta)\dot{\theta}^2 + a\cos(\theta)\ddot{\theta} - b\sin(\alpha)\dot{\alpha}^2 + b\cos(\alpha)\ddot{\alpha} \\
& \quad -c\sin(\beta)\dot{\beta}^2 + c\cos(\beta)\ddot{\beta} = 0.
\end{aligned} \tag{5.24}
$$

Again, one can write the Eq. 5.24 using the Jacobian matrix, obtaining

$$
\begin{bmatrix} b\sin(\alpha) & c\sin(\beta) \\ b\cos(\alpha) & c\cos(\beta) \end{bmatrix} \begin{Bmatrix} \ddot{\alpha} \\ \ddot{\beta} \end{Bmatrix} = \begin{Bmatrix} -a\cos(\theta)\dot{\theta}^2 - a\sin(\theta)\ddot{\theta} - b\cos(\alpha)\dot{\alpha}^2 \\ a\sin(\theta)\dot{\theta}^2 - a\cos(\theta)\ddot{\theta} + b\sin(\alpha)\dot{\alpha}^2 \\ -c\cos(\beta)\dot{\beta}^2 \\ +c\sin(\beta)\dot{\beta}^2 \end{Bmatrix}. \tag{5.25}
$$

Point of interest position analysis

For the point of interest analysis, consider the coupler point as shown in Fig. 5.18.

Point P position relative to the fixed referential is given by

$$
\begin{aligned}
x_p &= a\cos(\theta) + x_{1p}\cos(\alpha) - y_{1p}\sin(\alpha) \\
y_p &= a\sin(\theta) + x_{1p}\sin(\alpha) + y_{1p}\cos(\alpha).
\end{aligned} \tag{5.26}
$$

One can write the Eq. 5.26 using matrix

$$
\begin{Bmatrix} x_p \\ y_p \end{Bmatrix} = \begin{Bmatrix} a\cos(\theta) \\ a\sin(\theta) \end{Bmatrix} + \begin{bmatrix} \cos(\alpha) & -\sin(\alpha) \\ \sin(\alpha) & \cos(\alpha) \end{bmatrix} \begin{Bmatrix} x_{1p} \\ y_{1p} \end{Bmatrix}. \tag{5.27}
$$

Point of interest velocity analysis
Deriving Eq. 5.26 with respect to time, one obtains the velocity of the point of interest

$$\begin{aligned} \dot{x_p} &= -a\sin(\theta)\dot{\theta} - x_{1p}\sin(\alpha)\dot{\alpha} - y_{1p}\cos(\alpha)\dot{\alpha} \\ \dot{y_p} &= \ \ a\cos(\theta)\dot{\theta} + x_{1p}\cos(\alpha)\dot{\alpha} - y_{1p}\sin(\alpha)\dot{\alpha} \end{aligned} \tag{5.28}$$

Point of interest acceleration analysis
Deriving Eq. 5.28 with respect to time, one obtains the acceleration of the point of interest

$$\begin{aligned} \ddot{x_p} &= -a\cos(\theta)\dot{\theta}^2 - a\sin(\theta)\ddot{\theta} - x_{1p}\cos(\alpha)\dot{\alpha}^2 - x_{1p}\sin(\alpha)\ddot{\alpha} \\ &\quad + y_{1p}\sin(\alpha)\dot{\alpha}^2 - y_{1p}\cos(\alpha)\ddot{\alpha} \\ \ddot{y_p} &= -a\sin(\theta)\dot{\theta}^2 + a\cos(\theta)\ddot{\theta} - x_{1p}\sin(\alpha)\dot{\alpha}^2 + x_{1p}\cos(\alpha)\ddot{\alpha} \\ &\quad - y_{1p}\cos(\alpha)\dot{\alpha}^2 - y_{1p}\sin(\alpha)\ddot{\alpha} \end{aligned} \tag{5.29}$$

5.6.2 *Mechanism with Two Loops*

The slider-crank mechanism and the four-bar mechanism, analyzed previously, have only one loop. To analyze the kinematics of a mechanism with two or more loops, each loop must be traversed independently. To better understand the need to go through each loop, an analogy can be made with electronic circuits.

In electronics, *Kirchhoff's laws* are used to determine the current and voltage in circuits.

Cut-set Law: at any node, the sum of the currents entering and leaving the node is zero.

Circulation Law: in any mesh, the algebraic sum of the differences in voltages is zero.

The equivalent laws in mechanics would be:

Static Law: in any kinematic pair, the sum of the forces acting on the pair and that the pair applies is zero. This law is equivalent to static analysis.

Kinematic Law: in any loop, the algebraic sum of the velocities is zero. This law is equivalent to kinematic analysis.

The mechanism shown in Fig. 5.19 is used in presses, other similar mechanisms can be seen in [4]. From a rotating θ input one can generate large forces on the piston (f link). It can be seen that this mechanism features two loops.

Position analysis—loop 1
Loop 1 in Fig. 5.19 is a four-bar mechanism composed of the a, b, c, and d links. The kinematic analysis of the four-bar mechanism has been done in Sect. 5.6.1, however the example in Fig. 5.19 differs in the reference of the angles α and β, so the position equations are given by

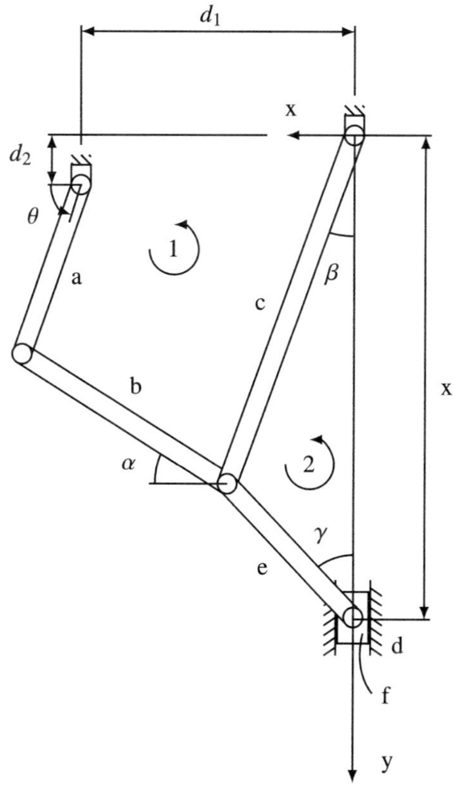

Fig. 5.19 Two-loop planar mechanism

$$\begin{aligned} x: &\quad d_1 + a\cos(\theta) - b\cos(\alpha) - c\sin(\beta) = 0 \\ y: &\quad d_2 + a\sin(\theta) + b\sin(\alpha) - c\cos(\beta) = 0. \end{aligned} \tag{5.30}$$

Velocity analysis—loop 1

Deriving Eq. 5.30 with respect to time, one has the velocity kinematics for loop 1, given by

$$\begin{aligned} x: &\quad -a\sin(\theta)\dot{\theta} + b\sin(\alpha)\dot{\alpha} - c\cos(\beta)\dot{\beta} = 0 \\ y: &\quad a\cos(\theta)\dot{\theta} + b\cos(\alpha)\dot{\alpha} + c\sin(\beta)\dot{\beta} = 0. \end{aligned} \tag{5.31}$$

Rearranging the Eq. 5.31 and using matrix, the Jacobian matrix can be seen.

$$\begin{bmatrix} b\sin(\alpha) & -c\cos(\beta) \\ b\cos(\alpha) & c\sin(\beta) \end{bmatrix} \begin{Bmatrix} \dot{\alpha} \\ \dot{\beta} \end{Bmatrix} = \begin{Bmatrix} a\sin(\theta)\dot{\theta} \\ -a\cos(\theta)\dot{\theta} \end{Bmatrix}. \tag{5.32}$$

Acceleration analysis—loop 1

Deriving Eq. 5.31 with respect to time, one has the acceleration kinematics for loop 1, given by

$$
\begin{aligned}
x: \quad & -a\cos(\theta)\dot{\theta}^2 - a\sin(\theta)\ddot{\theta} + b\cos(\alpha)\dot{\alpha}^2 + b\sin(\alpha)\ddot{\alpha} \\
& +c\sin(\beta)\dot{\beta}^2 - c\cos(\beta)\ddot{\beta} = 0 \\
y: \quad & -a\sin(\theta)\dot{\theta}^2 + a\cos(\theta)\ddot{\theta} - b\sin(\alpha)\dot{\alpha}^2 + b\cos(\alpha)\ddot{\alpha} \\
& +c\cos(\beta)\dot{\beta}^2 + c\sin(\beta)\ddot{\beta} = 0.
\end{aligned} \tag{5.33}
$$

Rearranging the Eq. 5.33 and using matrix, the Jacobian matrix can be seen.

$$
\begin{bmatrix} b\sin(\alpha) & -c\cos(\beta) \\ b\cos(\alpha) & c\sin(\beta) \end{bmatrix} \begin{Bmatrix} \ddot{\alpha} \\ \ddot{\beta} \end{Bmatrix} = \begin{Bmatrix} a\cos(\theta)\dot{\theta}^2 + a\sin(\theta)\ddot{\theta} \\ a\sin(\theta)\dot{\theta}^2 - a\cos(\theta)\ddot{\theta} \\ -b\cos(\alpha)\dot{\alpha}^2 - c\sin(\beta)\dot{\beta}^2 \\ +b\sin(\alpha)\dot{\alpha}^2 - c\cos(\beta)\dot{\beta}^2 \end{Bmatrix}. \tag{5.34}
$$

Position analysis—loop 2

Loop 2 in Fig. 5.19 is a slider-crank mechanism composed of the c, e, f, and d links. The kinematic analysis of the slider-crank was developed in the Sects. 5.1, 5.2 and 5.3, thus the expressions for the kinematic analysis are presented below, considering the nomenclature adopted in Fig. 5.19. The position analysis is given by

$$
\begin{aligned}
x: \quad & c\sin(\beta) - e\sin(\gamma) = 0 \\
y: \quad & c\cos(\beta) + e\cos(\gamma) - x = 0.
\end{aligned} \tag{5.35}
$$

Isolating the variables x and γ, it gives

$$
\begin{aligned}
\gamma &= \arcsin\left(\frac{c\sin(\beta)}{e}\right) \\
x &= c\cos(\beta) + e\cos(\gamma).
\end{aligned} \tag{5.36}
$$

Velocity analysis—loop 2

Deriving Eq. 5.35 with respect to time, one has the velocity kinematics for loop 2, given by

$$
\begin{aligned}
x: \quad & c\cos(\beta)\dot{\beta} - e\cos(\gamma)\dot{\gamma} = 0 \\
y: \quad & -c\sin(\beta)\dot{\beta} - e\sin(\gamma)\dot{\gamma} - \dot{x} = 0.
\end{aligned} \tag{5.37}
$$

Rearranging to make the Jacobian matrix explicit

$$
\begin{bmatrix} e\cos(\gamma) & 0 \\ e\sin(\gamma) & 1 \end{bmatrix} \begin{Bmatrix} \dot{\gamma} \\ \dot{x} \end{Bmatrix} = \begin{Bmatrix} c\cos(\beta)\dot{\beta} \\ -c\sin(\beta)\dot{\beta} \end{Bmatrix}. \tag{5.38}
$$

Isolating the desired variables, $\dot{x}$ and $\dot{\gamma}$, we have

$$
\begin{aligned}
\dot{\gamma} &= \frac{c\cos(\beta)\dot{\beta}}{e\cos(\gamma)} \\
\dot{x} &= -c\sin(\beta)\dot{\beta} - e\sin(\gamma)\dot{\gamma}.
\end{aligned} \tag{5.39}
$$

Acceleration analysis—loop 2

Deriving the Eq. 5.37 with respect to time, one has the acceleration kinematics for loop 2, given by

$$\begin{aligned} x: &\quad -c\sin(\beta)\dot{\beta}^2 + c\cos(\beta)\ddot{\beta} + e\sin(\gamma)\dot{\gamma}^2 - e\cos(\gamma)\ddot{\gamma} &= 0 \\ y: &\quad -c\cos(\beta)\dot{\beta}^2 - c\sin(\beta)\ddot{\beta} - e\cos(\gamma)\dot{\gamma}^2 - e\sin(\gamma)\ddot{\gamma} - \ddot{x} &= 0. \end{aligned} \tag{5.40}$$

Rearranging to make the Jacobian matrix explicit

$$\begin{bmatrix} e\cos(\gamma) & 0 \\ e\sin(\gamma) & 1 \end{bmatrix} \begin{Bmatrix} \ddot{\gamma} \\ \ddot{x} \end{Bmatrix} = \begin{Bmatrix} -c\sin(\beta)\dot{\beta}^2 + c\cos(\beta)\ddot{\beta} + e\sin(\gamma)\dot{\gamma}^2 \\ -c\cos(\beta)\dot{\beta}^2 - c\sin(\beta)\ddot{\beta} - e\cos(\gamma)\dot{\gamma}^2 \end{Bmatrix}. \tag{5.41}$$

Isolating the desired variables, $\ddot{x}$ and $\ddot{\gamma}$, we have

$$\begin{aligned} \ddot{\gamma} &= \frac{-c\sin(\beta)\dot{\beta}^2 + c\cos(\beta)\ddot{\beta} + e\sin(\gamma)\dot{\gamma}^2}{e\cos(\gamma)} \\ \ddot{x} &= -c\cos(\beta)\dot{\beta}^2 - c\sin(\beta)\ddot{\beta} - e\cos(\gamma)\dot{\gamma}^2 - e\sin(\gamma)\ddot{\gamma}. \end{aligned} \tag{5.42}$$

Point of interest analysis

For a press, the point of interest is the piston which executes the pressing. Thus, we analyze the position of the piston which is given by x, $\dot{x}$ and $\ddot{x}$, in Eqs. 5.36, 5.39 and 5.42, respectively. Thus, solving the kinematics equations determines the position, velocity and acceleration of the piston for a given position, velocity and acceleration of the motor. However, before solving the equations for the point of interest you must solve the equations for loops 1 and 2.

5.6.3 Mechanism with Two Degrees of Freedom

The mechanism in Fig. 5.20 has mobility 2. Thus, to fully position the mechanism, it is necessary to define the angles θ_1 and θ_2.

Position analysis

Analyzing the loop in x and y, one has the position equations

$$\begin{aligned} x: &\quad a\cos(\theta_1) + b\cos(\alpha) - c\cos(\theta_2) - d &= 0 \\ y: &\quad a\sin(\theta_1) + b\sin(\alpha) - c\sin(\theta_2) &= 0 \end{aligned} \tag{5.43}$$

Notice in Eq. 5.43 that it is possible to obtain an analytical solution. To do so, simply isolate the variable b in one of the equations and replace it in the other equation. Isolating b in the x equation

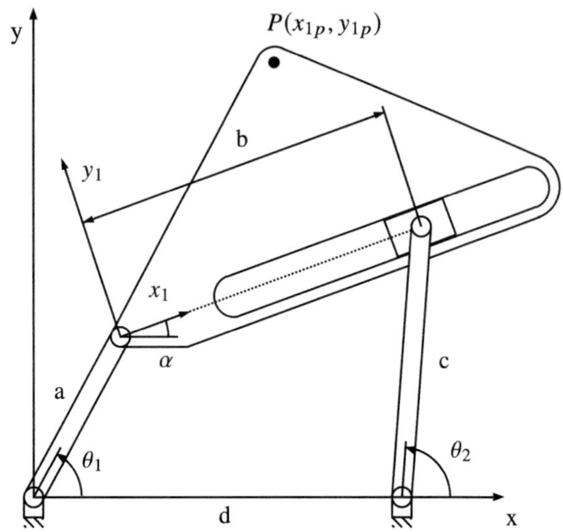

Fig. 5.20 Planar mechanism with two degrees of freedom

$$a\cos(\theta_1) + b\cos(\alpha) - c\cos(\theta_2) - d = 0$$

$$-b\cos(\alpha) = a\cos(\theta_1) - c\cos(\theta_2) - d$$

$$b = \frac{-a\cos(\theta_1) + c\cos(\theta_2) + d}{\cos(\alpha)}$$

Substituting b into the y equation

$$a\sin(\theta_1) + \frac{-a\cos(\theta_1) + c\cos(\theta_2) + d}{\cos(\alpha)}\sin(\alpha) - c\sin(\theta_2) = 0$$

$$a\sin(\theta_1) + (-a\cos(\theta_1) + c\cos(\theta_2) + d)\tan(\alpha) - c\sin(\theta_2) = 0$$

$$\alpha = \arctan\left(\frac{c\sin(\theta_2) - a\sin(\theta_1)}{-a\cos(\theta_1) + c\cos(\theta_2) + d}\right).$$

Velocity analysis
Deriving the Eq. 5.43 with respect to time, one has

$$\begin{aligned} x: &\quad -a\sin(\theta_1)\dot{\theta}_1 + \dot{b}\cos(\alpha) - b\sin(\alpha)\dot{\alpha} + c\sin(\theta_2)\dot{\theta}_2 = 0 \\ y: &\quad a\cos(\theta_1)\dot{\theta}_1 + \dot{b}\sin(\alpha) + b\cos(\alpha)\dot{\alpha} - c\cos(\theta_2)\dot{\theta}_2 = 0. \end{aligned} \tag{5.44}$$

Evidencing the Jacobian matrix

$$\begin{bmatrix} \cos(\alpha) & -b\sin(\alpha) \\ \sin(\alpha) & +b\cos(\alpha) \end{bmatrix} \begin{Bmatrix} \dot{b} \\ \dot{\alpha} \end{Bmatrix} = \begin{Bmatrix} a\sin(\theta_1)\dot{\theta}_1 - c\sin(\theta_2)\dot{\theta}_2 \\ -a\cos(\theta_1)\dot{\theta}_1 + c\cos(\theta_2)\dot{\theta}_2 \end{Bmatrix}. \tag{5.45}$$

To determine $\dot{b}$ and $\dot{\alpha}$, one can isolate $\dot{b}$ from Eq. 5.44 for x:

$$\dot{b} = \frac{a\sin(\theta_1)\dot{\theta}_1 + b\sin(\alpha)\dot{\alpha} - c\sin(\theta_2)\dot{\theta}_2}{\cos(\alpha)}. \tag{5.46}$$

Substituting Eq. 5.46 into Eq. 5.44 for y:

$$a\cos(\theta_1)\dot{\theta}_1 + \frac{a\sin(\theta_1)\dot{\theta}_1 + b\sin(\alpha)\dot{\alpha} - c\sin(\theta_2)\dot{\theta}_2}{\cos(\alpha)}\sin(\alpha) + b\cos(\alpha)\dot{\alpha} \\ -c\cos(\theta_2)\dot{\theta}_2 = 0$$

$$a\cos(\theta_1)\dot{\theta}_1 + (a\sin(\theta_1)\dot{\theta}_1 + b\sin(\alpha)\dot{\alpha} - c\sin(\theta_2)\dot{\theta}_2)\tan(\alpha) \\ +b\cos(\alpha)\dot{\alpha} - c\cos(\theta_2)\dot{\theta}_2 = 0$$

Isolating $\dot{\alpha}$

$$\dot{\alpha}(b\sin(\alpha)\tan(\alpha) + b\cos(\alpha)) + a\cos(\theta_1)\dot{\theta}_1 + (a\sin(\theta_1)\dot{\theta}_1 \\ -c\sin(\theta_2)\dot{\theta}_2)\tan(\alpha) - c\cos(\theta_2)\dot{\theta}_2 = 0$$

Finally, we have $\dot{\alpha}$

$$\dot{\alpha} = \frac{-a\cos(\theta_1)\dot{\theta}_1 - (a\sin(\theta_1)\dot{\theta}_1 - c\sin(\theta_2)\dot{\theta}_2)\tan(\alpha) + c\cos(\theta_2)\dot{\theta}_2}{b\sin(\alpha)\tan(\alpha) + b\cos(\alpha)}. \tag{5.47}$$

Acceleration analysis
Deriving the Eq. 5.44 with respect to time, one has

$$\begin{aligned} x: \quad & -a\cos(\theta_1)\dot{\theta}_1^2 - a\sin(\theta_1)\ddot{\theta}_1 + \ddot{b}\cos(\alpha) - \dot{b}\sin(\alpha)\dot{\alpha} - \dot{b}\sin(\alpha)\dot{\alpha} \\ & -b\cos(\alpha)\dot{\alpha}^2 - b\sin(\alpha)\ddot{\alpha} + c\cos(\theta_2)\dot{\theta}_2^2 + c\sin(\theta_2)\ddot{\theta}_2 = 0 \\ y: \quad & -a\sin(\theta_1)\dot{\theta}_1^2 + a\cos(\theta_1)\ddot{\theta}_1 + \ddot{b}\sin(\alpha) + \dot{b}\cos(\alpha)\dot{\alpha} + \dot{b}\cos(\alpha)\dot{\alpha} \\ & -b\sin(\alpha)\dot{\alpha}^2 + b\cos(\alpha)\ddot{\alpha} + c\sin(\theta_2)\dot{\theta}_2^2 - c\cos(\theta_2)\ddot{\theta}_2 = 0 \end{aligned}$$

$$\begin{aligned} x: \quad & -a\cos(\theta_1)\dot{\theta}_1^2 - a\sin(\theta_1)\ddot{\theta}_1 + \ddot{b}\cos(\alpha) - 2\dot{b}\sin(\alpha)\dot{\alpha} \\ & -b\cos(\alpha)\dot{\alpha}^2 - b\sin(\alpha)\ddot{\alpha} + c\cos(\theta_2)\dot{\theta}_2^2 + c\sin(\theta_2)\ddot{\theta}_2 = 0 \\ y: \quad & -a\sin(\theta_1)\dot{\theta}_1^2 + a\cos(\theta_1)\ddot{\theta}_1 + \ddot{b}\sin(\alpha) + 2\dot{b}\cos(\alpha)\dot{\alpha} \\ & -b\sin(\alpha)\dot{\alpha}^2 + b\cos(\alpha)\ddot{\alpha} + c\sin(\theta_2)\dot{\theta}_2^2 - c\cos(\theta_2)\ddot{\theta}_2 = 0. \end{aligned} \tag{5.48}$$

Evidencing the Jacobian matrix

$$\begin{bmatrix} \cos(\alpha) & -b\sin(\alpha) \\ \sin(\alpha) & +b\cos(\alpha) \end{bmatrix} \begin{Bmatrix} \ddot{b} \\ \ddot{\alpha} \end{Bmatrix} = \left\{ \begin{matrix} a\cos(\theta_1)\dot{\theta_1}^2 + a\sin(\theta_1)\ddot{\theta_1} \\ a\sin(\theta_1)\dot{\theta_1}^2 - a\cos(\theta_1)\ddot{\theta_1} \end{matrix} \right. \\ \left. \begin{matrix} +2\dot{b}\sin(\alpha)\dot{\alpha} + b\cos(\alpha)\dot{\alpha}^2 - c\cos(\theta_2)\dot{\theta_2}^2 - c\sin(\theta_2)\ddot{\theta_2} \\ -2\dot{b}\cos(\alpha)\dot{\alpha} + b\sin(\alpha)\dot{\alpha}^2 - c\sin(\theta_2)\dot{\theta_2}^2 + c\cos(\theta_2)\ddot{\theta_2} \end{matrix} \right\}. \tag{5.49}$$

To determine $\ddot{b}$ and $\ddot{\alpha}$, one can isolate $\ddot{b}$ from Eq. 5.48 for x:

$$\ddot{b} = \frac{a\cos(\theta_1)\dot{\theta_1}^2 + a\sin(\theta_1)\ddot{\theta_1} + 2\dot{b}\sin(\alpha)\dot{\alpha} + b\cos(\alpha)\dot{\alpha}^2}{\cos(\alpha)} \\ + \frac{b\sin(\alpha)\ddot{\alpha} - c\cos(\theta_2)\dot{\theta_2}^2 - c\sin(\theta_2)\ddot{\theta_2}}{\cos(\alpha)}. \tag{5.50}$$

Substituting Eq. 5.50 into Eq. 5.48 for y, one has

$$-a\sin(\theta_1)\dot{\theta_1}^2 + a\cos(\theta_1)\ddot{\theta_1} + \sin(\alpha)\frac{a\cos(\theta_1)\dot{\theta_1}^2 + a\sin(\theta_1)\ddot{\theta_1}}{\cos(\alpha)} \\ + \sin(\alpha)\frac{2\dot{b}\sin(\alpha)\dot{\alpha} + b\cos(\alpha)\dot{\alpha}^2 + b\sin(\alpha)\ddot{\alpha} - c\cos(\theta_2)\dot{\theta_2}^2}{\cos(\alpha)} \\ - \sin(\alpha)\frac{c\sin(\theta_2)\ddot{\theta_2}}{\cos(\alpha)} + 2\dot{b}\cos(\alpha)\dot{\alpha} - b\sin(\alpha)\dot{\alpha}^2 + b\cos(\alpha)\ddot{\alpha} \\ + c\sin(\theta_2)\dot{\theta_2}^2 - c\cos(\theta_2)\ddot{\theta_2} = 0.$$

$$-a\sin(\theta_1)\dot{\theta_1}^2 + a\cos(\theta_1)\ddot{\theta_1} + \Big(a\cos(\theta_1)\dot{\theta_1}^2 + a\sin(\theta_1)\ddot{\theta_1} + 2\dot{b}\sin(\alpha)\dot{\alpha} \\ + b\cos(\alpha)\dot{\alpha}^2 + b\sin(\alpha)\ddot{\alpha} - c\cos(\theta_2)\dot{\theta_2}^2 - c\sin(\theta_2)\ddot{\theta_2}\Big)\tan(\alpha) \\ + 2\dot{b}\cos(\alpha)\dot{\alpha} - b\sin(\alpha)\dot{\alpha}^2 + b\cos(\alpha)\ddot{\alpha} + c\sin(\theta_2)\dot{\theta_2}^2 \\ - c\cos(\theta_2)\ddot{\theta_2} = 0.$$

Isolating $\ddot{\alpha}$

$$\ddot{\alpha}\,(b\sin(\alpha)\tan(\alpha) + b\cos(\alpha)) - a\sin(\theta_1)\dot{\theta_1}^2 + a\cos(\theta_1)\ddot{\theta_1} \\ + \tan(\alpha)\Big(a\cos(\theta_1)\dot{\theta_1}^2 + a\sin(\theta_1)\ddot{\theta_1} + 2\dot{b}\sin(\alpha)\dot{\alpha} + b\cos(\alpha)\dot{\alpha}^2 \\ - c\cos(\theta_2)\dot{\theta_2}^2 - c\sin(\theta_2)\ddot{\theta_2}\Big) + 2\dot{b}\cos(\alpha)\dot{\alpha} - b\sin(\alpha)\dot{\alpha}^2 + c\sin(\theta_2)\dot{\theta_2}^2 \\ - c\cos(\theta_2)\ddot{\theta_2} = 0.$$

Finally, we have $\ddot{\alpha}$:

$$\ddot{\alpha} = \frac{A}{B} \tag{5.51}$$

where

$$\begin{aligned} A = a\sin(\theta_1)\dot{\theta_1}^2 - a\cos(\theta_1)\ddot{\theta_1} - \tan(\alpha)\Big(a\cos(\theta_1)\dot{\theta_1}^2 + a\sin(\theta_1)\ddot{\theta_1} \\ +2\dot{b}\sin(\alpha)\dot{\alpha} + b\cos(\alpha)\dot{\alpha}^2 - c\cos(\theta_2)\dot{\theta_2}^2 - c\sin(\theta_2)\ddot{\theta_2}\Big) - 2\dot{b}\cos(\alpha)\dot{\alpha} \\ +b\sin(\alpha)\dot{\alpha}^2 - c\sin(\theta_2)\dot{\theta_2}^2 + c\cos(\theta_2)\ddot{\theta_2} \end{aligned} \tag{5.52}$$

$$B = b\sin(\alpha)\tan(\alpha) + b\cos(\alpha). \tag{5.53}$$

Point of interest position analysis
For the point of interest analysis, consider the point P at Fig. 5.20. The position of point P in relation to the fixed coordinated system is given by

$$\begin{aligned} x_p &= a\cos(\theta_1) + x_{1p}\cos(\alpha) - y_{1p}\sin(\alpha) \\ y_p &= a\sin(\theta_1) + x_{1p}\sin(\alpha) + y_{1p}\cos(\alpha). \end{aligned} \tag{5.54}$$

Rewriting Eq. 5.54 to explicit the rotation matrix, yields

$$\begin{Bmatrix} x_p \\ y_p \end{Bmatrix} = \begin{Bmatrix} a\cos(\theta_1) \\ a\sin(\theta_1) \end{Bmatrix} + \begin{bmatrix} \cos(\alpha) & -\sin(\alpha) \\ \sin(\alpha) & \cos(\alpha) \end{bmatrix} \begin{Bmatrix} x_{1p} \\ y_{1p} \end{Bmatrix}. \tag{5.55}$$

Point of interest velocity analysis
Deriving the Eq. 5.54 with respect to time, one has the point of interest velocity

$$\begin{aligned} \dot{x_p} &= -a\sin(\theta_1)\dot{\theta_1} - x_{1p}\sin(\alpha)\dot{\alpha} - y_{1p}\cos(\alpha)\dot{\alpha} \\ \dot{y_p} &= \ \ a\cos(\theta_1)\dot{\theta_1} + x_{1p}\cos(\alpha)\dot{\alpha} - y_{1p}\sin(\alpha)\dot{\alpha}. \end{aligned} \tag{5.56}$$

Point of interest acceleration analysis
Deriving the Eq. 5.56 with respect to time, one has the point of interest acceleration

$$\begin{aligned} \ddot{x_p} = -a\cos(\theta_1)\dot{\theta_1}^2 - a\sin(\theta_1)\ddot{\theta_1} - x_{1p}\cos(\alpha)\dot{\alpha}^2 - x_{1p}\sin(\alpha)\ddot{\alpha} \\ +y_{1p}\sin(\alpha)\dot{\alpha}^2 - y_{1p}\cos(\alpha)\ddot{\alpha} \\ \ddot{y_p} = -a\sin(\theta_1)\dot{\theta_1}^2 + a\cos(\theta_1)\ddot{\theta_1} - x_{1p}\sin(\alpha)\dot{\alpha}^2 + x_{1p}\cos(\alpha)\ddot{\alpha} \\ -y_{1p}\cos(\alpha)\dot{\alpha}^2 - y_{1p}\sin(\alpha)\ddot{\alpha}. \end{aligned} \tag{5.57}$$

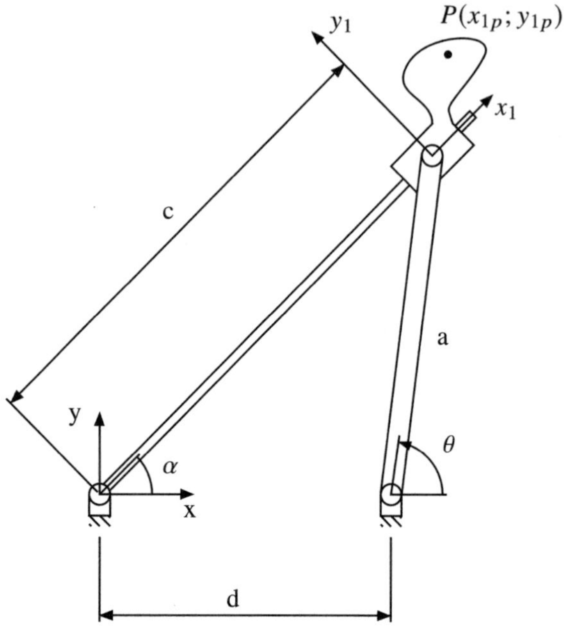

Fig. 5.21 Planar slider-crank inversion

Problems

5.1 Do the position, velocity, and acceleration kinematic analysis for the slider-crank inversion shown in Fig. 5.21. Also determine the position, velocity, and acceleration equations for the indicated point of interest. The input variable is θ, and the variables to be determined are α and c.

5.2 The rapid return mechanism goal is to move from a start point to an end point at a given speed and then return to the start point as quickly as possible. Do the position, velocity, and acceleration kinematic analysis for the rapid return mechanism shown in Fig. 5.22. The input variable is θ and the variables to be determined are α and x. Hint: the mechanism has two loops, and c is a dependent variable of θ.

5.3 The slider-crank mechanism has had a link and a kinematic pair added, so d_1 is a second input variable. Do the position, velocity, and acceleration kinematic analysis for the mechanism shown in Fig. 5.23. Also, determine the position, velocity, and acceleration equations for the indicated point of interest. The input variables are θ and d_1, and the variables to be determined are β and x.

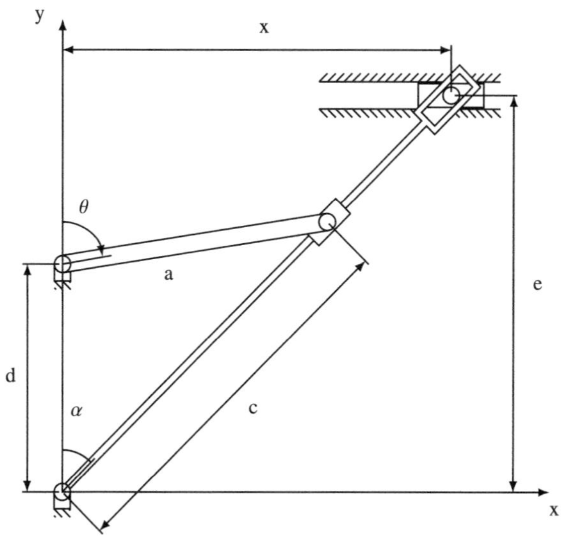

Fig. 5.22 Quick return planar mechanism

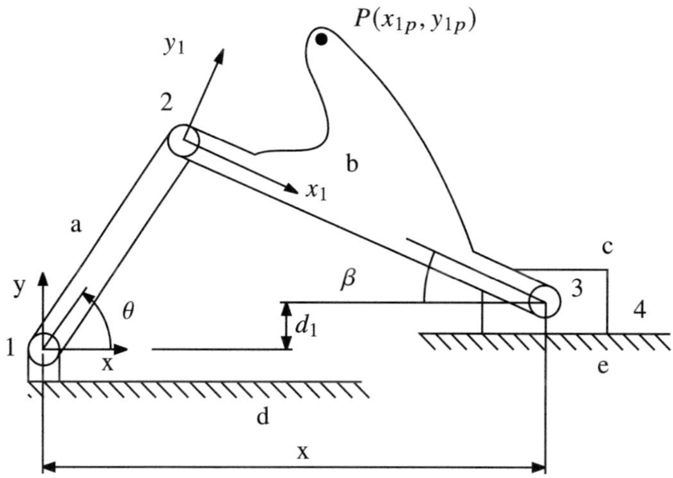

Fig. 5.23 Slider-crank planar mechanism with added link and pair

References

1. Norton, R.L.: Design of Machinery: An Introduction to the Synthesis and Analysis of Mechanisms and Machines. McGraw-Hill, New York (2008)
2. McCarthy, J., Soh, G.: Geometric Design of Linkages. Springer, New York (2010)
3. Suh, C., Radcliffe, C.: Kinematics and Mechanisms Design. R.E. Krieger Pub. Co., Malabar (1978)
4. Hydraulic and mechanical presses. In: The Library of Manufacturing (2014). Available at: http://thelibraryofmanufacturing.com/presses.html

Part III
Mechanism Synthesis

Chapter 6
Number and Type Syntheses

Abstract This chapter introduces the systematic synthesis of mechanisms, focusing on two stages: number and type syntheses. It starts with defining structural and design requirements and explains how these criteria guide the selection and configuration of links and joints. Number synthesis is detailed through combinatorial approaches to determine how many links and joints are required to achieve a specified function. Type synthesis then explores selecting appropriate kinematic pairs—such as revolute, prismatic, or cam pairs—to fulfill the required motion within those configurations. Practical examples and exercises reinforce the enumeration and classification of feasible mechanisms. The chapter concludes by situating number and type syntheses within the broader design methodology, linking them to preliminary and conceptual design stages. By establishing a structured approach to mechanism configuration, this chapter helps students and engineers systematically transition from a functional requirement to a viable mechanical layout.

Keywords Number synthesis · Type synthesis · Dimensional synthesis · Kinematic chain enumeration

This chapter covers the synthesis of mechanisms. Initially, an overview of mechanism synthesis, its stages and what each stage comprises is covered. Then, number synthesis is looked at in detail and with practical examples. Finally, some concepts about type synthesis is covered.

6.1 Number, Type and Dimensional Syntheses

As seen in Chap. 1, synthesis is the process of obtaining the mechanism geometry from the design specifications. These specifications can be divided into two categories.

Structural requirements: specifications that determine or involve the mechanism structural characteristics. Examples of structural requirements are: mobility 2, non-complex mechanism (few independent loops) and planar workspace ($\lambda = 3$).

D. Martins and E. H. Murai, *Mechanisms–Synthesis and Analysis with Applications in Robotics*, Mechanisms and Machine Science 183,
https://doi.org/10.1007/978-3-031-95187-9_6

Design requirements: general specifications involving non-measurable properties or measurable properties that are not structural characteristics of mechanisms. Examples of design requirements are: lightness, ergonomics and compactness (because the structure of the mechanism does not take its dimensions into account).

The focus of this text is the synthesis of mechanisms, not machines or products. Therefore, in general, only structural requirements are used. The design requirements would mainly be used in stages after the mechanism synthesis, stages that is not covered in this book.

Mechanism synthesis can be divided into three stages [1].

Number synthesis: aims to determine the **number** of links and kinematic pairs in the mechanism as well as which links are connected to each other.

Type synthesis: aims to determine which are the **types** of kinematic pairs that connect the links, e.g. revolute pair, cylindrical pair, prismatic pair, gear, cam.

Dimensional synthesis: aims to determine the **dimensions** of the links (lengths and angles) so that the mechanism has the desired kinematics.

The number synthesis is seen in more detail in Sect. 6.3, the type synthesis in Sect. 6.4, and the dimensional synthesis is only shown for the four-bar mechanism. Kinematic design[1] of the four-bar mechanism is covered in Chaps. 7, 8 and 9.

6.2 Mechanism Design Methodology

Mechanism design methodologies organize the steps described in Sect. 6.1 in order to systematize the mechanism design process. There are several mechanism design methodologies, as can be seen in [1–3]. Each of these methodologies has its own characteristics and differences, but in general their structures are similar. This section presents a methodology that covers the most general stages of mechanism design. Finally, compared to design methodologies, mechanism design methodologies are more specific, so their stages are seen in more detail than those of design methodologies. However, there are similarities between both methodologies, which are shown at the end of this section. The steps presented in the Sect. 6.1 are organized as shown in Fig. 6.1.

The procedure begins by surveying the state of the art. The aim of this stage is to list all existing devices that have a similar function to the one desired. Another objective is to gain a better understanding of the problem to be solved, observing which characteristics have made one or more solutions better or more feasible than others.

[1] The term "kinematic design" is used to differentiate it from "structural design", which deals with stress analysis. The term "kinematic" will be omitted. Thus, the term "design" will always refer to kinematic design when it is not accompanied by a specification.

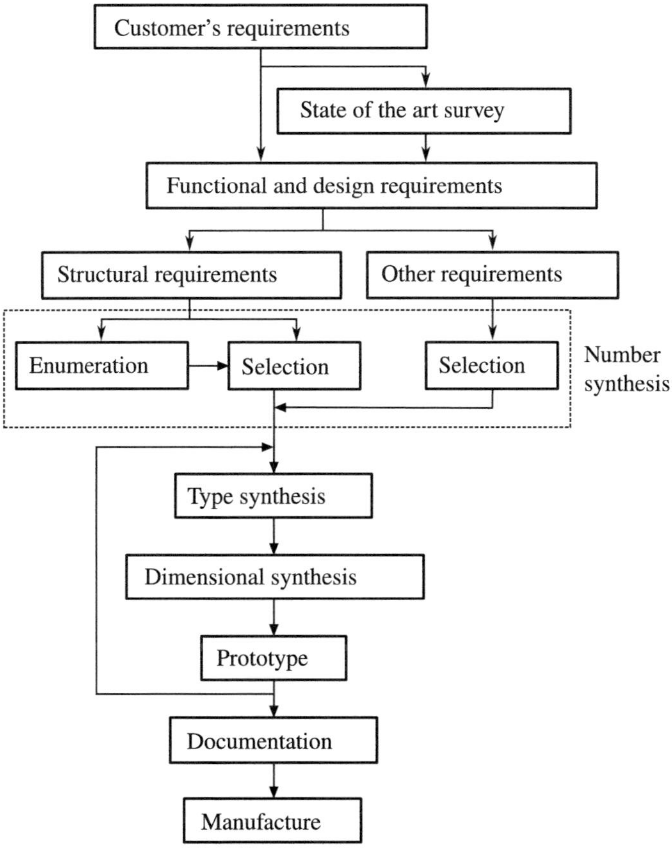

Fig. 6.1 Mechanism design methodology

The mechanisms found in the state-of-the-art survey should be analyzed. This step will help with future project decision-making, such as deciding which design and structural requirements are the most relevant.

The structural and design requirements must then be determined. The structural requirements are used in the number synthesis both to generate kinematic chains and to identify feasible solutions.

Number synthesis has already been described in the Sect. 6.1. Number synthesis uses some structural requirements to enumerate kinematic chains and mechanisms. Other structural requirements are used to evaluate the solutions generated and identify feasible solutions. Thus, at this stage synthesis is combined with analysis in order to generate feasible mechanisms. If there are no feasible results, the structural requirements adopted initially should be re-evaluated.

For the feasible mechanisms generated in the number synthesis, the type synthesis determines the type of each kinematic pair. Next is dimensional synthesis, which determines the dimensions of each component of the mechanisms.

Once the dimensions have been determined, a prototype of the mechanism can be made to evaluate its operation. Figure 6.1 shows that these steps are contained within a *loop*, so that any adjustments can be made until an optimum solution is obtained. Once the mechanism has been determined, it is documented and then manufactured.

It should be noted that more steps could be included in the methodology shown in Fig. 6.1, but the aim of this figure is to provide an overview of the mechanism synthesis process without going into detail.

Finally, there are similarities between the methodology presented here and design methodologies in general. Thus, the stages from surveying the state of the art to determining the project requirements would be contained within the informational stage. The number synthesis stage and, in some cases, part of the type synthesis stage would be contained within the conceptual design and preliminary design. Dimensional synthesis would be contained within the detailed design stage. Other stages are also often found individualized in design methodologies.

6.3 Number Synthesis

The number synthesis starts from a set of *structural requirements* to determine the number of links, kinematic pairs and which links are connected to each other, obtaining various mechanisms that satisfy the structural requirements. The reader is encouraged to review the concept of mobility, seen in Chap. 2, and isomorphism, seen in Sect. 4.4.

6.3.1 Basic Concepts for Number Synthesis

Some basic concepts are necessary to understand number synthesis, namely partition, variation and inversion.

Partition is the set of all links used to assemble the kinematic chain. So a partition can have unary, binary, ternary links and so on. In mechanism synthesis, the kinematic pair elements must be distributed between the links. Thus, if a link receives three kinematic pair elements it is a ternary link, if it receives four elements it is quaternary. Another point of view is that, in the end, the sum of the number of elements in all the links results in e. Thus, a partition is nothing more than a way of dividing the number of elements. This point of view is interesting because it refers to the mathematical problem of partitioning.[2]

[2] In number theory, in mathematics, the problem of partitioning aims to determine all possible ways of dividing an integer into a sum of positive integers. The nomenclature and also the tools used in this stage of number synthesis come from this branch of mathematics.

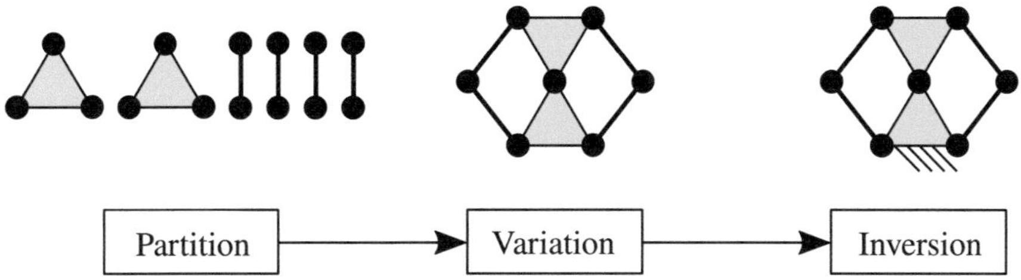

Fig. 6.2 Partition, variation and inversion sequence

Variation[3] (kinematic chains) is the way in which the links of a *partition* can be arranged to generate a kinematic chain. Thus, given a partition, connecting the links in all possible ways generates all possible variations. Note that determining all possible variations is equivalent to determining all possible kinematic chains that can be assembled for the partition in analysis.

Inversion (mechanisms) is to associate a fixed reference to one of the links in a kinematic chain. Thus, starting from a kinematic chain, determining all its inversions is equivalent to determining all mechanisms that can be generated.

The number synthesis sequence presented above is illustrated in Fig. 6.2.

6.3.2 Procedure for Number Synthesis

Number synthesis starts from a set of structural requirements. Usually, the structural requirements are mobility (M), the dimension of the workspace (λ) and the number of independent loops (ν). With these three structural requirements, one can use the *mobility equation* (Eq. 2.4) and the *Euler equation* (Eq. 2.5) to determine the number of links (n) and the number of kinematic pairs with pair connectivity equals to 1 (j). With j determined, the number of kinematic pair elements e can be calculated.

Once the number of links and kinematic pair elements has been calculated, all possible *partitions* are determined. Once one has all possible partitions, for each partition it is determined all possible *variations* (kinematic chains). Once all possible kinematic chains have been determined, all possible *inversions* (mechanisms) can be obtained for each chain.

A summary of the number synthesis is shown below:

1. Determine the number of links (n), the number of kinematic pairs with unit pair connectivity (j) and the number of kinematic pair elements (e) for the given structural requirements.
2. Determine all possible partitions to be generated for n and e calculated.
3. Determine all possible variations for each partition obtained.
4. Determine all possible inversions for each variation obtained.

[3] Not to be confused with the concept of *variety* seen in Sect. 4.1.

The number synthesis problem is a combinatorial analysis problem and, as such, the number of results tends to grow very quickly when increasing the structural requirements values. Determining the structural requirements must be done with caution, not only to avoid extra workload on the synthesis, but also because the requirements quality directly influences the solution quality.

6.3.3 Challenges of Number Synthesis

Number synthesis presents some challenges. In the number synthesis process, some kinematic chains and mechanisms can present certain characteristics that are not desired. These unwanted characteristics are divided into two types: **degeneration** and **isomorphism**.

Degenerate (or improper) kinematic chains are not considered possible solutions. For a kinematic chain to be considered degenerate, it must have one of the following characteristics:

- Baranov subchain;
- Fractionation.

Baranov subchain is a subchain whose mobility is zero. Thus, a part of the kinematic chain has no degrees of freedom and it can be considered a structure.

Baranov subchains can be generated during the process of joining the links to generate the kinematic chain. As the subchain is considered a structure, all the links and kinematic pairs that make up the chain can be replaced by a single link. Therefore, its structural characteristics, such as the number of links, kinematic pairs and loops, differ from those calculated for the structural requirements. Therefore, a kinematic chain with a Baranov subchain does not meet the structural requirements. An example of a kinematic chain with a Baranov subchain is shown in Fig. 6.3. Two examples of Baranov structures are shown in Fig. 6.4.

Due to the differences in their structural characteristics, kinematic chains with Baranov subchains should be identified and eliminated from the enumeration process.

Fractionation occurs when the removal of a link or the opening (separation) of a kinematic pair results in two disconnected kinematic chains. An example of fractionation is shown in Fig. 6.5.

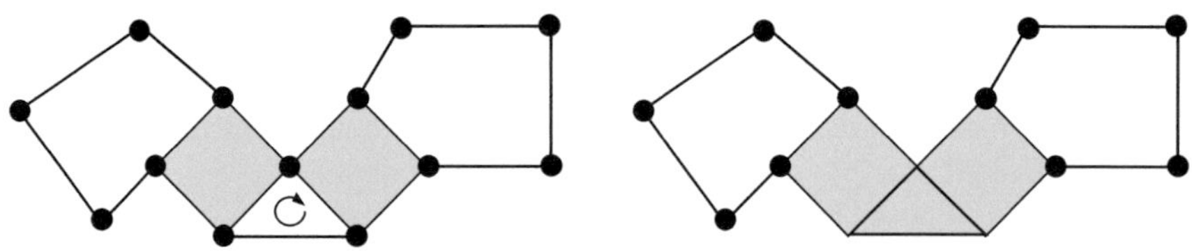

Fig. 6.3 Baranov's subchain in a kinematic chain

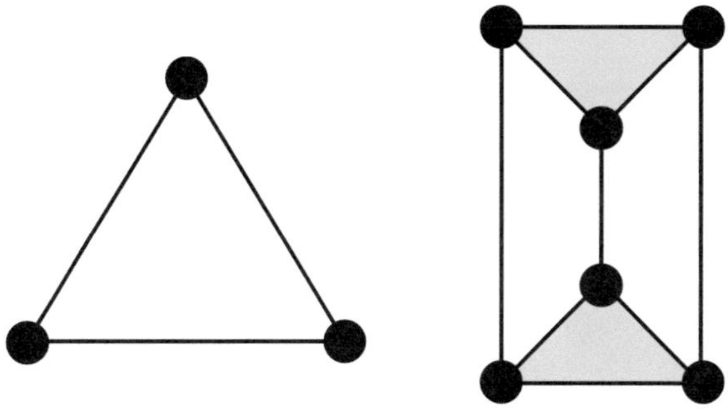

Fig. 6.4 Examples of Baranov's subchains

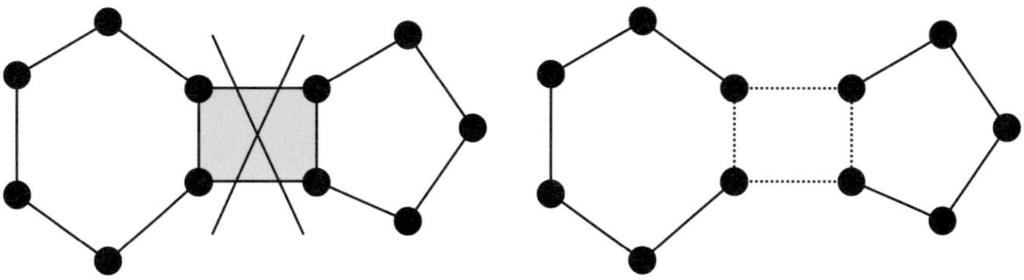

Fig. 6.5 Link fractionation

Fractionation can be classified into three types.

Link fractionation: occurs when, by removing a link from the kinematic chain, it splits into two disconnected parts.

Kinematic pair fractionation: occurs when, by disconnecting the elements of a kinematic pair, the kinematic chain splits into two disconnected parts.

Fractionation in hybrid chains: is characterized by the sequential occurrence of the two previous fractionations.

Kinematic chains that exhibit fractionation are not original solutions, but merely the union of two kinematic chains. In addition, all fractionated kinematic pairs must have actuators. This imposes contraints on the design and it can make fractionated kinematic chains unfeasible. However, in certain circumstances, such a solution may be valid. Therefore, fractionation is not always considered a degeneration and must be analyzed on a case-by-case basis.

If fractionation is considered a degeneration, kinematic chains that have fractionation must be identified and eliminated from the enumeration process.

The concept of *isomorphism* has already been introduced in Sect. 4.4. When two kinematic chains or mechanisms are isomorphic, they have the same structural characteristics. The difference between such chains or mechanisms is in the links numbering.

Figure 6.6 shows two Watt mechanisms that are isomorphic to each other. From one of them you can get the other just by renumbering the links.

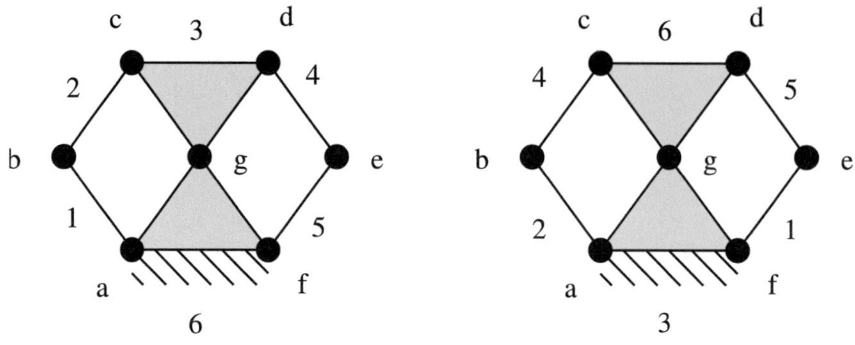

Fig. 6.6 Isomorphism in Watt's mechanism

An isomorphic kinematic chain does not generate a different solution and should be eliminated from the synthesis process. The same applies to isomorphic mechanisms.

As we have seen, the number synthesis process is a combinatorial analysis. Thus, carry on the synthesis process with a degenerate or isomorphous kinematic chain entails a large computational cost to arrive at results that are not solutions or are not new solutions. Therefore, degenerate and isomorphic kinematic chains should be eliminated from the synthesis process as soon as possible.

Number synthesis

Determine all mechanisms with a closed kinematic chain that satisfy the following structural requirements: $M = 1$, $\nu = 2$ and $\lambda = 3$ (planar space).

Solution:

(I) Determining the number of links and kinematic pairs

Using Eq. 2.4 and Eq. 2.5, calculate the number of links and the number of kinematic pairs with unit pair connectivity. Obtaining $n = 6$ and $j = 7$

The mechanism should have six links and seven kinematic pairs with unit pair connectivity. Each kinematic pair is the union of two kinematic pair elements. Thus, the number of kinematic pair elements is given by

$$e = 2j. \tag{6.1}$$

Obtaining $e = 14$.

(II) Determining the partitions

The 14 kinematic pair elements must be distributed among the six links. As the statement calls for the kinematic chain to be closed, every link must have at least two kinematic pair elements. Distributing two kinematic pair elements to each link leaves $\bar{e}$, with $\bar{e} = e - 2.n$ elements to be distributed.

The partitions can be determined by determining all possible ways of distributing the $\bar{e}$ elements.

$$\bar{e} = e - 2n = 14 - 2 * 6 = 2$$

There are two ways to distribute the remaining two elements:

Mode 1: $2 \rightarrow 2$
Mode 2: $2 \rightarrow 1 + 1$

Adding the kinematic pair elements partitioned above with the two kinematic pair elements that each link must contain, we obtain the partitions for $M = 1$, $\nu = 2$ and $\lambda = 3$:

Partition 1: $4 + 2 + 2 + 2 + 2 + 2$
Partition 2: $3 + 3 + 2 + 2 + 2 + 2$

(III) **Determining the variations**

In the first partition it is not possible to make proper kinematic chains, since any arrangement generates a Baranov subchain. In the second partition, there are two non-isomorphic chains: the Watt chain, shown in Fig. 6.7a, and the Stephenson chain, shown in Fig. 6.7b.

(IV) **Determining the inversions**
The non-isomorphic kinematic inversions for the two kinematic chains are shown in Fig. 6.8.

Number synthesis

Determine all mechanisms with a closed kinematic chain that satisfy the following structural requirements: $M = 1$, $\nu = 3$ and $\lambda = 3$ (planar space).

Fig. 6.7 Kinematic chains for partition 2

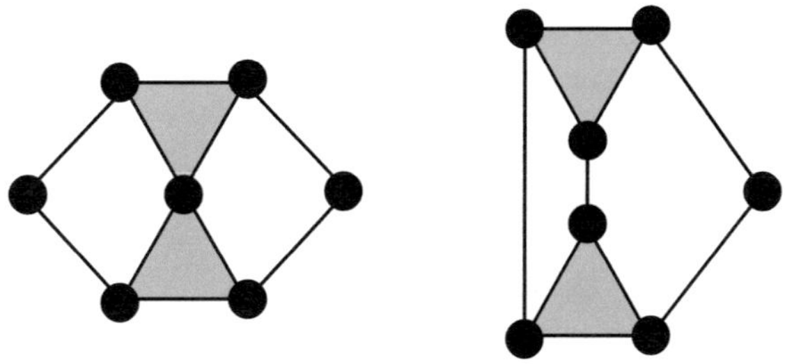

(a) (b)

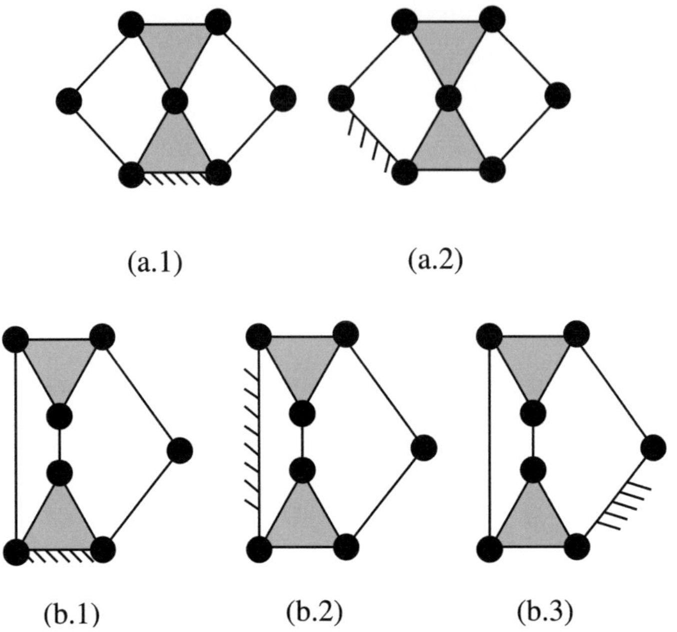

Fig. 6.8 Mechanisms for partition 2

Solution:

(I) Determining the number of links and kinematic pairs

Using Eq. 2.4 and Eq. 2.5, calculate the number of links and the number of kinematic pairs with unit pair connectivity. Obtaining $n = 8$ and $j = 10$.

The mechanism should have eight links and 10 kinematic pairs with unit pair connectivity. Each kinematic pair is the union of two kinematic pair elements. Thus, the number of kinematic pair elements is given by

$$e = 2j. \tag{6.2}$$

Obtaining $e = 20$.

(II) Determining the partitions

The 20 kinematic pair elements must be distributed among the eight links. As the statement requires the kinematic chain to be closed, every link must have at least two kinematic pair elements. Distributing two kinematic pair elements to each link leaves $\bar{e}$, where $\bar{e} = e - 2n$ are the elements to be distributed.

All partitions are determined by determining all possible ways of distributing the $\bar{e}$ elements.

$$\bar{e} = e - 2n = 20 - 2 * 8 = 4$$

Distributing the remaining four elements in all possible ways:

Mode 1: $4 \rightarrow 4$
Mode 2: $4 \rightarrow 3 + 1$
Mode 3: $4 \rightarrow 2 + 2$
Mode 4: $4 \rightarrow 2 + 1 + 1$
Mode 5: $4 \rightarrow 1 + 1 + 1 + 1$

Adding the two kinematic pair elements that had already been distributed to each link, we have the partitions:

Partition 1: $6 + 2 + 2 + 2 + 2 + 2 + 2 + 2$
Partition 2: $5 + 3 + 2 + 2 + 2 + 2 + 2 + 2$
Partition 3: $4 + 4 + 2 + 2 + 2 + 2 + 2 + 2$
Partition 4: $4 + 3 + 3 + 2 + 2 + 2 + 2 + 2$
Partition 5: $3 + 3 + 3 + 3 + 2 + 2 + 2 + 2$

Figure 6.9 graphically shows the procedure for determining the partitions. Thus, the first partition contains one hexanary link and seven binary links. The second partition contains one quinary link, one ternary link and six binary links. The third partition contains two quaternary links and six binary links. The fourth partition contains one quaternary link, two ternary links and five binary links. The fifth partition contains four ternary links and four binary links. Any other distribution of kinematic pair elements results in a partition equal to or equivalent to the five partitions that have been determined.

(III) **Determining the variations**

For partition 1, it is not possible to set up any proper kinematic chains. All results have Baranov subchains or disconnected kinematic pair elements.

For partition 2, it is not possible to set up any proper kinematic chains. All results have Baranov subchains or disconnected kinematic pair elements.

For partition 3, two proper kinematic chains are possible, as shown in Fig. 6.10. Any other kinematic chain assembled from this partition has a Baranov subchain or is isomorphic to one of the two chains shown in Fig. 6.10.

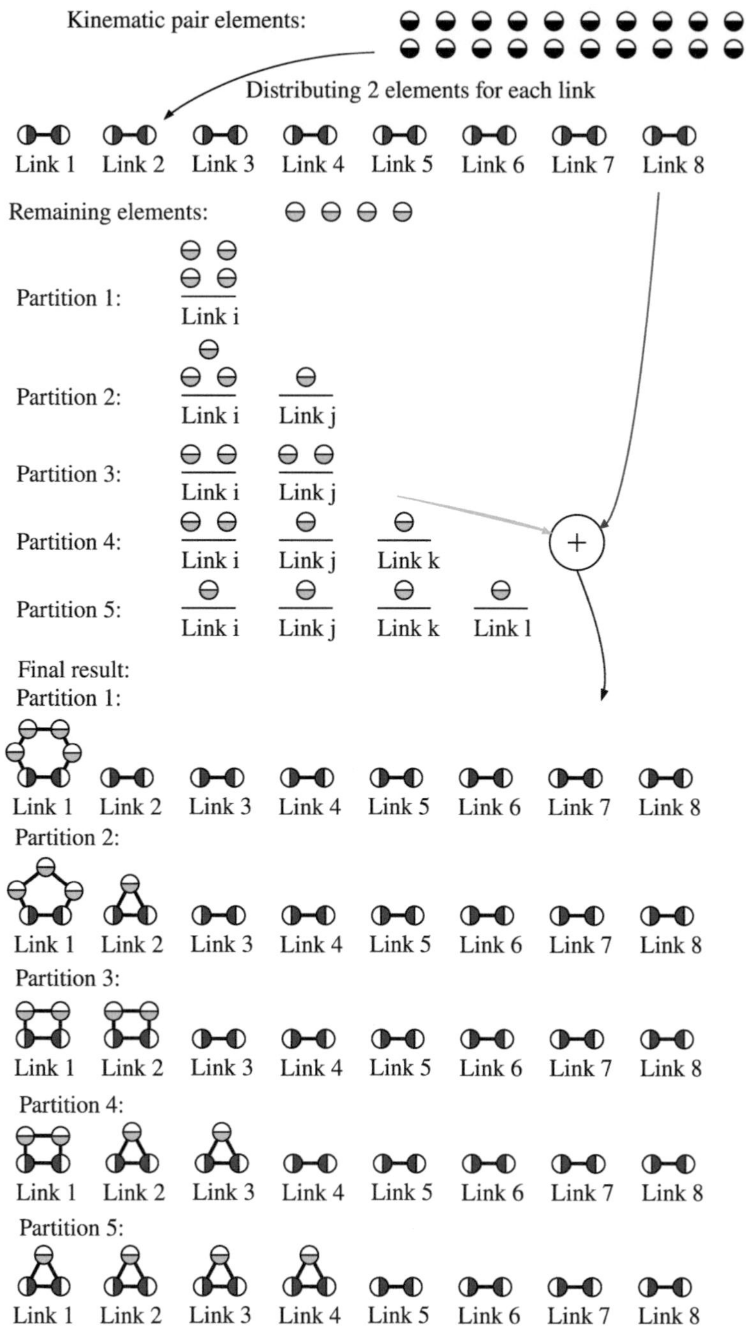

Fig. 6.9 Determining the partitions

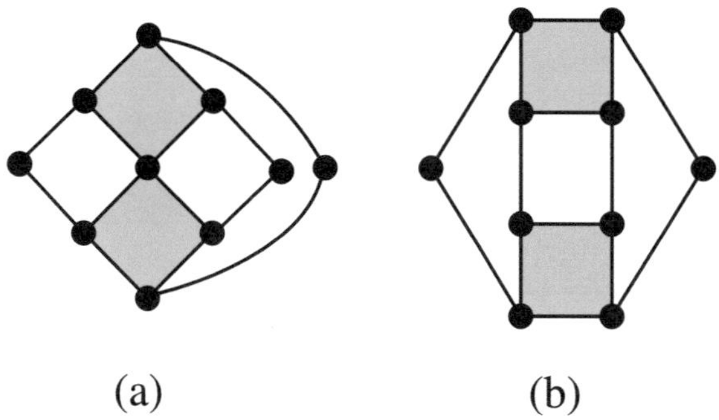

Fig. 6.10 Possible variations for partition 3

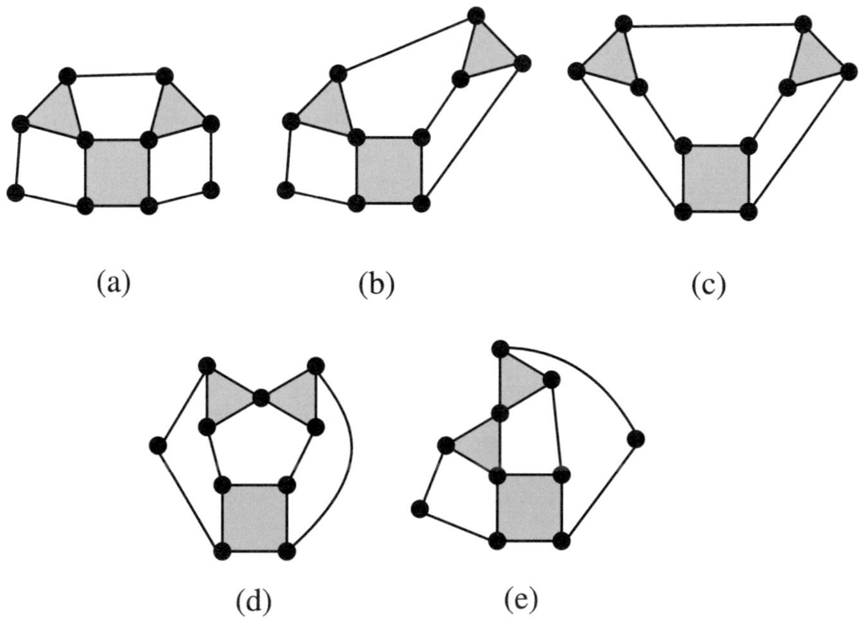

Fig. 6.11 Possible variations for partition 4

For partition 4, five proper kinematic chains are possible, as shown in Fig. 6.11. Any other kinematic chain assembled from this partition has a Baranov subchain or is isomorphic to one of the five chains shown in Fig. 6.11.

For partition 5, nine proper kinematic chains are possible, as shown in Fig. 6.12. Any other kinematic chain has a Baranov subchain or is isomorphic to one of the nine chains shown in Fig. 6.12.

(IV) **Determining the inversions**

For each kinematic chain in Figs. 6.10, 6.11 and 6.12, a link must be chosen to be the fixed link, generating a mechanism.

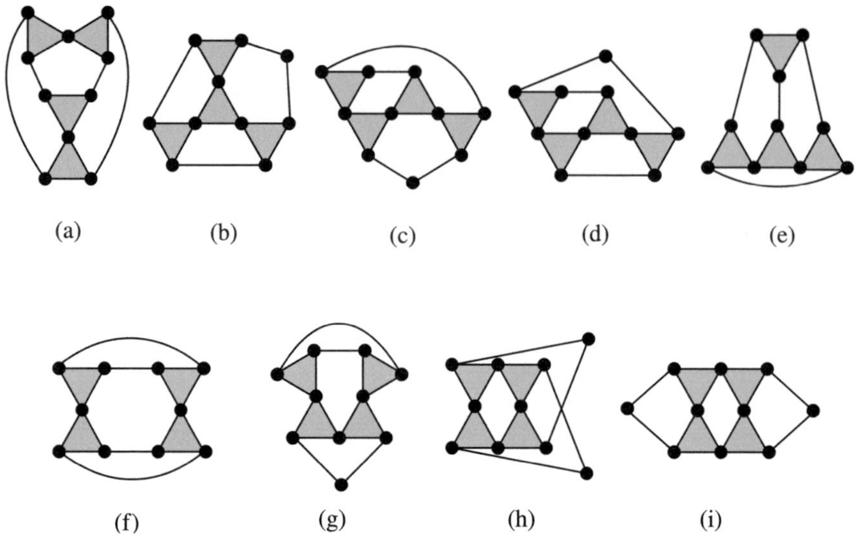

Fig. 6.12 Possible variations for partition 5

Table 6.1 shows how many non-isomorphic mechanisms can be generated for each kinematic chain.

Table 6.2 shows the number of planar kinematic chains (only unfractionated) for some values of M and λ.

For more information on the number synthesis, see [1–5].

6.4 Type Synthesis

The **type synthesis** aims to determine the type of each kinematic pair in a mechanism. Initially, all types of kinematic pairs that can be used must be listed. Factors that can make it impossible to use some types of kinematic pair are:

Workspace: kinematic pairs must meet the workspace constraints. For example, if the workspace is planar, revolute, prismatic and planar pairs can be used, as well as higher pairs of gears and cams. A counterexample would be the use of a cylindrical pair, since it allows motions that are not contained in the planar space.

Manufacturing: available manufacturing processes can affect the choice of available or economically feasible pairs.

Maintenance: the mechanism must be easy to maintain and, in the event of failure, the pairs must be easy to replace. The use of a superior pair tends to be more critical because it requires more maintenance. If a non-commercial pair is used, such as cams and some types of gears, stock must be kept for replacement, resulting in additional costs.

Table 6.1 All the 71 non-isomorphic mechanisms generated by partitions 3, 4 and 5

Partition	Kinematic chain	Possible inversions	Mechanism
3	a	2	See Figs. 6.13a.1 and 6.13a.2.
3	b	3	See Figs. 6.13b.1 to 6.13b.3.
4	a	5	See Figs. 6.14a.1 to 6.14a.5.
4	b	7	See Figs. 6.14b.1 to 6.14b.7.
4	c	4	See Figs. 6.14c.1 to 6.14c.4.
4	d	7	See Figs. 6.14d.1 to 6.14d.7.
4	e	8	See Figs. 6.15e.1 to 6.15e.8.
5	a	2	See Figs. 6.16a.1 and 6.16a.2.
5	b	5	See Figs. 6.16b.1 to 6.16b.5.
5	c	8	See Figs. 6.16c.1 to 6.17c.8.
5	d	4	See Figs. 6.17d.1 to 6.17d.4.
5	e	6	See Figs. 6.17e.1 to 6.17e.6.
5	f	2	See Figs. 6.17f.1 and 6.17f.2.
5	g	4	See Figs. 6.17g.1 to 6.17g.4.
5	h	2	See Figs. 6.17h.1 and 6.17h.2.
5	i	2	See Figs. 6.17i.1 and 6.17i.2.
		71	

Table 6.2 Number of planar kinematic chains, adapted from [4]

Mobility	1	2	3	4	5	6
$\nu = 2$	2	3	5	6	8	10
$\nu = 3$	16	35	74	126	212	325
$\nu = 4$	230	753	1962	4356	8846	16649
$\nu = 5$	6856	27496	38547	216291	–	–
$\nu = 6$	318162	1432730	4805764	13743920	–	–
$\nu = 7$	19819281	–	–	–	–	–

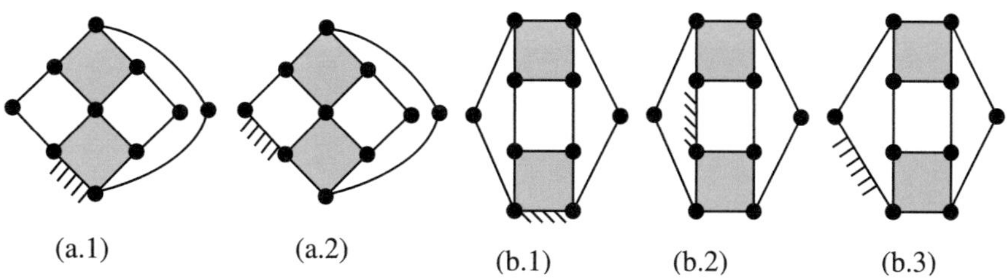

Fig. 6.13 Possible inversions for partition 3

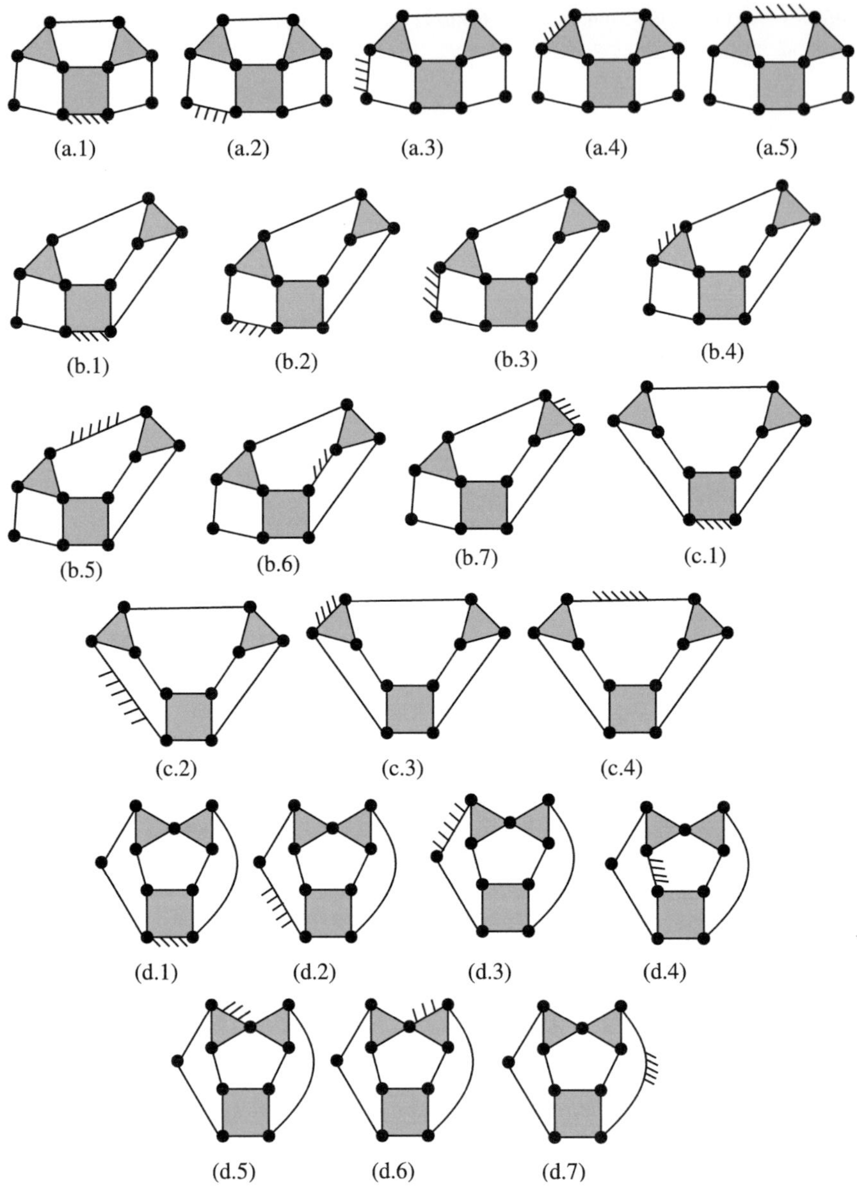

Fig. 6.14 Possible inversions for partition 4

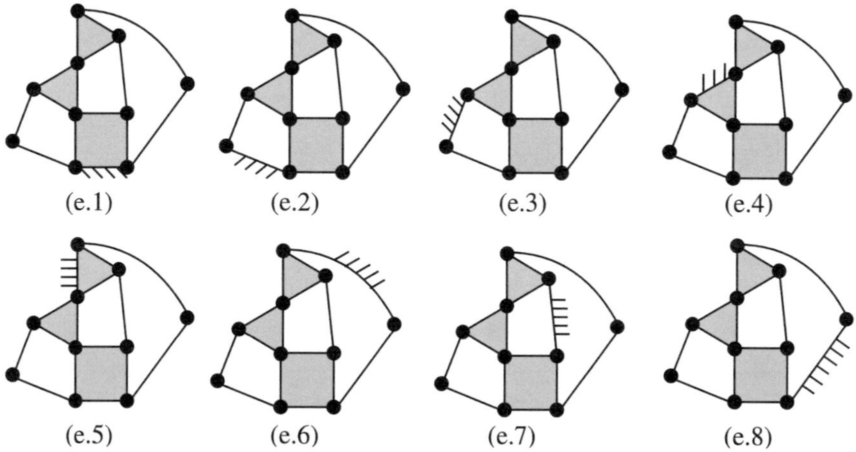

Fig. 6.15 Possible inversions for partition 4 (continued)

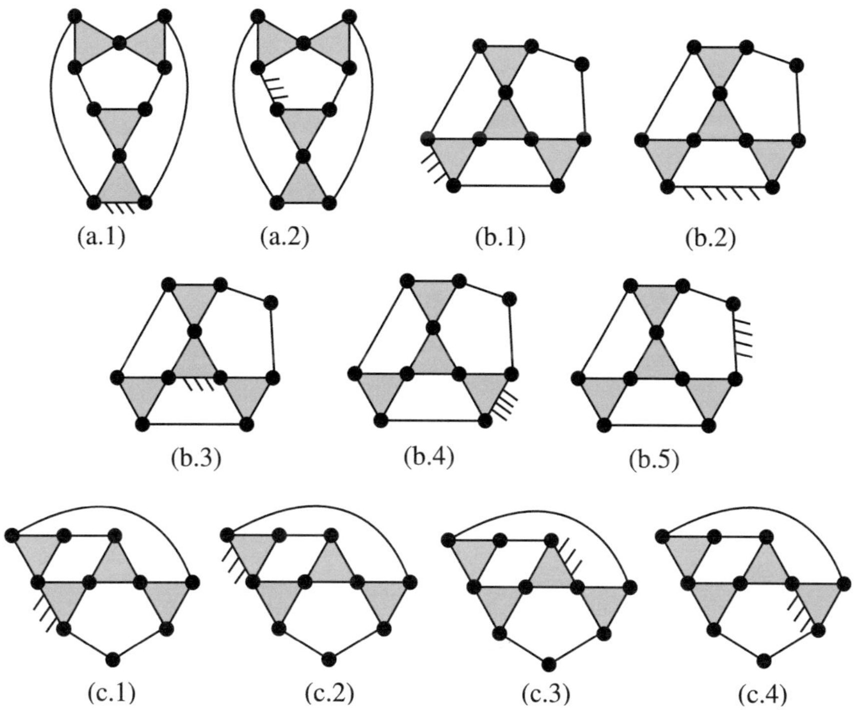

Fig. 6.16 Possible inversions for partition 5

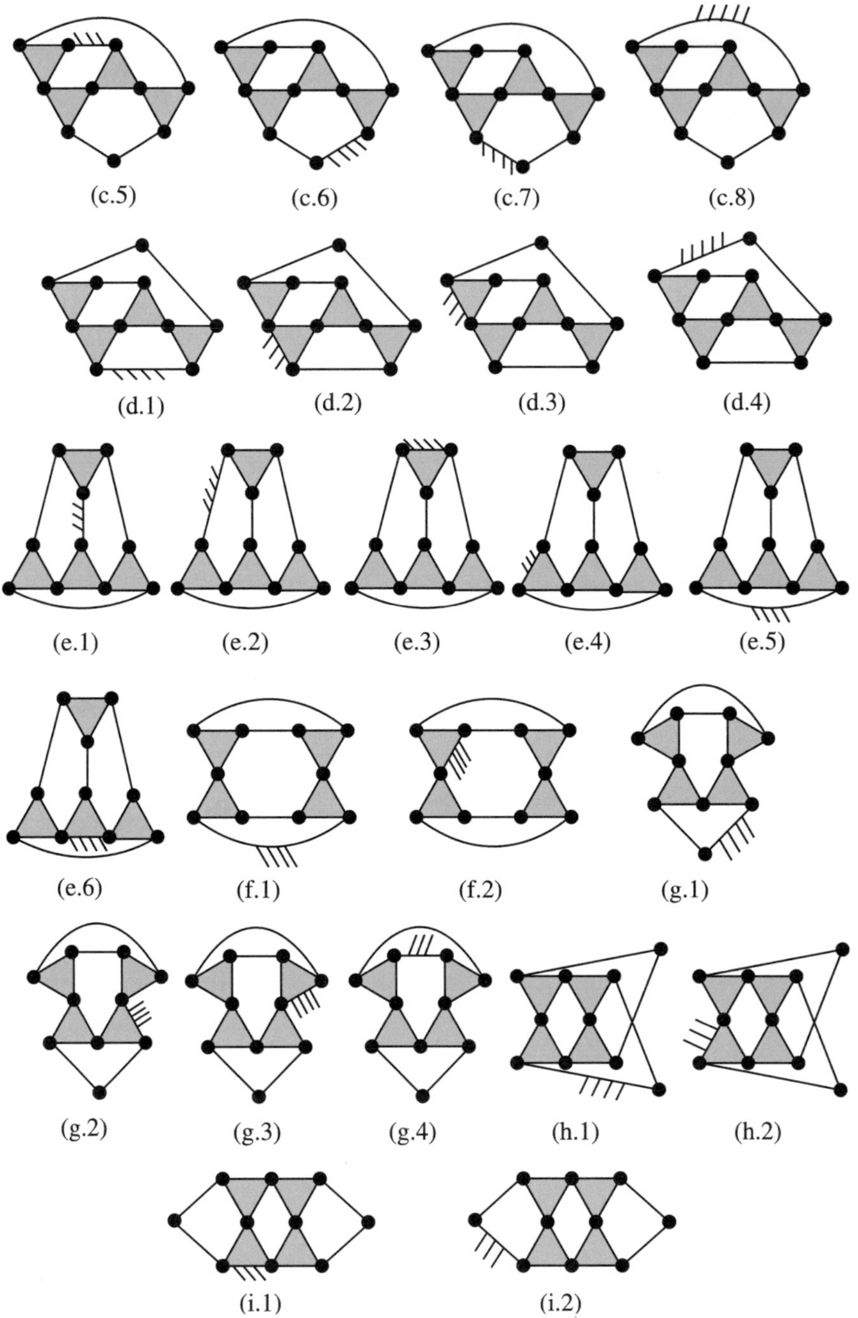

Fig. 6.17 Possible inversions for partition 5 (continued)

Note that the mechanisms resulting from Example 6.3.3 have generalized kinematic pairs with unit connectivity, i.e. each kinematic pair represents a degree of freedom regardless of the pair type. Thus, the mechanisms shown in Example 6.3.3 can have revolute pairs or prismatic, as these are the only planar pairs with pair connectivity equal to one (unitary). Note that the *kinematic pair contraction* can be used to combine two pairs with connectivity one into a pair with connectivity two, thus being able to use a gear or a cam.[4]

Removing the unfeasible kinematic pairs, for each mechanism we determine all possible mechanisms with the given pair types. This is done by assigning all possible pair types to the kinematic pairs. Like number synthesis, type synthesis is a combinatorial analysis problem in which the number of results grows very quickly as the number of kinematic pairs and types increases. It is, therefore, recommended to eliminate as many pair types as possible and use design requirements to eliminate unfeasible combinations.

An example of using design requirements to eliminate unfeasible solutions would be a requirement to use motors for actuation. In this case, the kinematic pairs to be actuated must have the rotational degree of freedom. Thus, there is one less pair to have the type determined, reducing the number of results.

Rear suspension synthesis for two-wheeled vehicles

Design a planar mechanism for the rear suspension of two-wheeled vehicles. Consider the following design requirements: $M = 1$, $\nu = 2$ and, since it is a planar linkage, $\lambda = 3$.

Mobility one is chosen because suspensions generally have only one degree of freedom, i.e. given a position for the wheel, the entire mechanism is completely positioned.

A single independent loop could have been chosen, resulting in a link pivoted in relation to the structure and with a shock absorber. There are many motorcycles and bicycles that use this suspension for its simplicity, low cost and weight. However, these suspensions have problems with efficiency, stability and control during braking and acceleration (*squat* and *bob* problems). The choice of three or more independent loops would result in more complex mechanisms, with more expensive manufacture and maintenance and greater weight. So $\nu = 2$ is chosen.

Solution:

The number synthesis in this case is the same as in Example 6.3.3, so the mechanisms obtained in Example 6.3.3 are repeated in Fig. 6.18.

[4] The reader will recall from the Sect. 2.1 that cams and gears have pair connectivity equal to two.

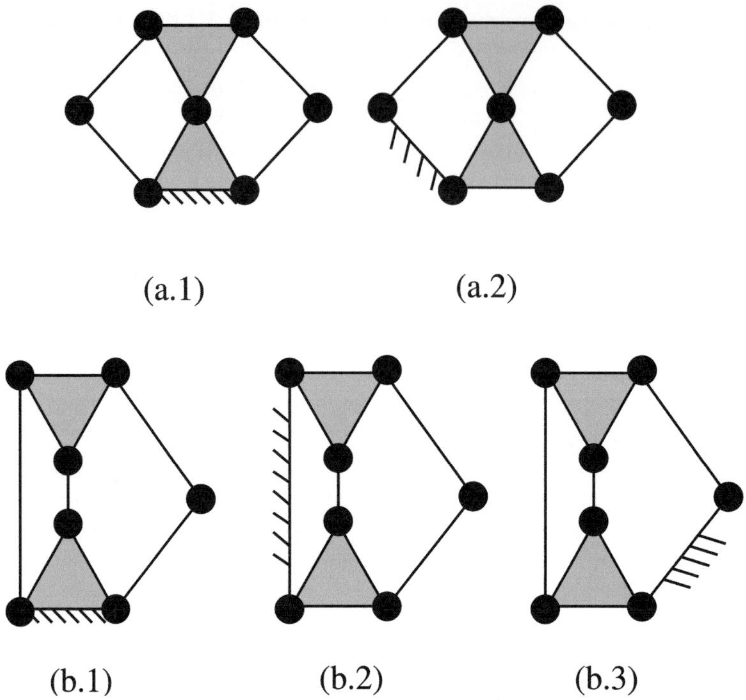

Fig. 6.18 Mechanisms for $M = 1$, $\nu = 2$ and $\lambda = 3$

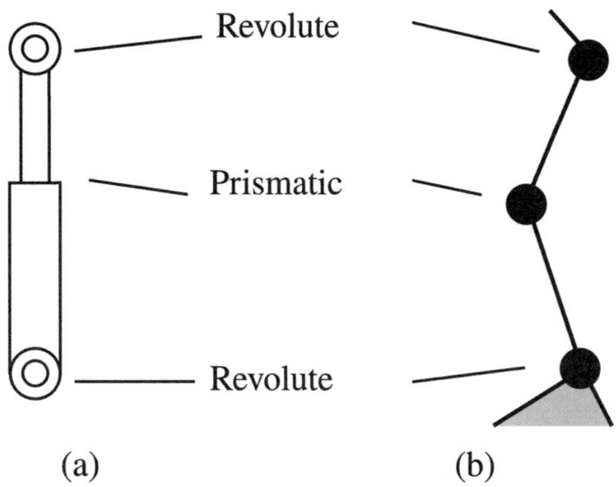

Fig. 6.19 Open subchain RPR of a shock absorber. **a** Ahock absorber. **b** Kinematic chain

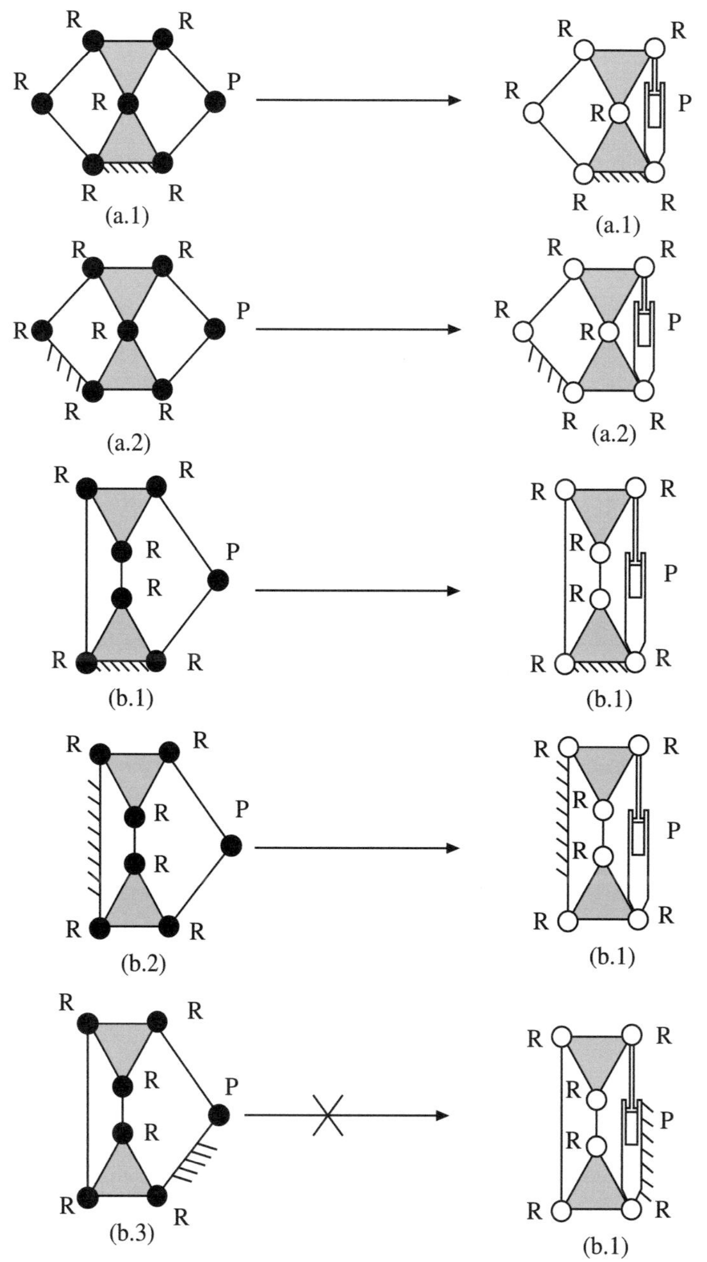

Fig. 6.20 Mechanisms for rear suspension

Type synthesis

The types of kinematic pairs available with unit pair connectivity and for the planar workspace are revolute pair and prismatic pair.

There are seven kinematic pairs and each pair has two options: prismatic and revolute. Thus, each mechanism in Fig. 6.8 has 2^7 mechanisms with the defined pair types. In other words, considering the five possible mechanisms, we have a total of:

$$5 * 2^7 = 640.$$

Thus, considering the structural requirements adopted, there are 640 possible mechanisms for the rear suspension of two-wheeled vehicles. Analyzing all 640 options is laborious, so design requirements must be used to eliminate unfeasible results.

The suspension's function is to absorb impacts in order to improve stability, control and comfort. Shock absorbers are used to do this. When inserting a shock absorber, the type of three kinematic pairs is defined, as shown in Fig. 6.19.

Thus, with three kinematic pairs defined, there are $5 * 2^4$ results left, i.e. 80 mechanisms (a considerable reduction from the original 640). For didactic purposes, consider that the 4 pairs to be defined in each mechanism are all revolute. In general, the only prismatic pair in suspensions is in the shock absorber, so this consideration is quite reasonable. This leaves five possible suspension mechanisms. The five mechanisms are shown in Fig. 6.20. It is suggested that the reader continue the analysis for the remaining 75 cases in which there is more than one prismatic kinematic pair.

Note that in the case of Fig. 6.20b.3, the shock absorber body would have to be attached to the motorcycle frame (as in front suspensions for two-wheeled vehicles). Another option would be to attach the shock absorber piston, which is also common in front suspensions. The same could happen in the mechanism in Fig. 6.20a.2. However, in rear suspensions this option is not usual. For didactic purposes, we will not consider this option. The remaining four mechanisms are shown in Fig. 6.21, where the practical application is easier to visualize.

Figure 6.22 shows the relationship between the mechanisms designed and suspensions found on motorcycles. Figure 6.22a shows the Kawasaki Uni-trak, Fig. 6.22b shows the Honda Unit Prolink suspension, Fig. 6.22c shows the Honda Prolink suspension and Fig. 6.22d shows the Suzuki Full Floater suspension. Bicycles also use similar suspensions, the most common being the Uni-trak type.

Finally, only kinematic chains with two independent loops have been considered, but it is a suggestion for the reader to do the type synthesis for motorcycle suspensions with $M = 1$, $\nu = 3$ and $\lambda = 3$. The number synthesis for these structural requirements has already been done in Example 6.3.3.

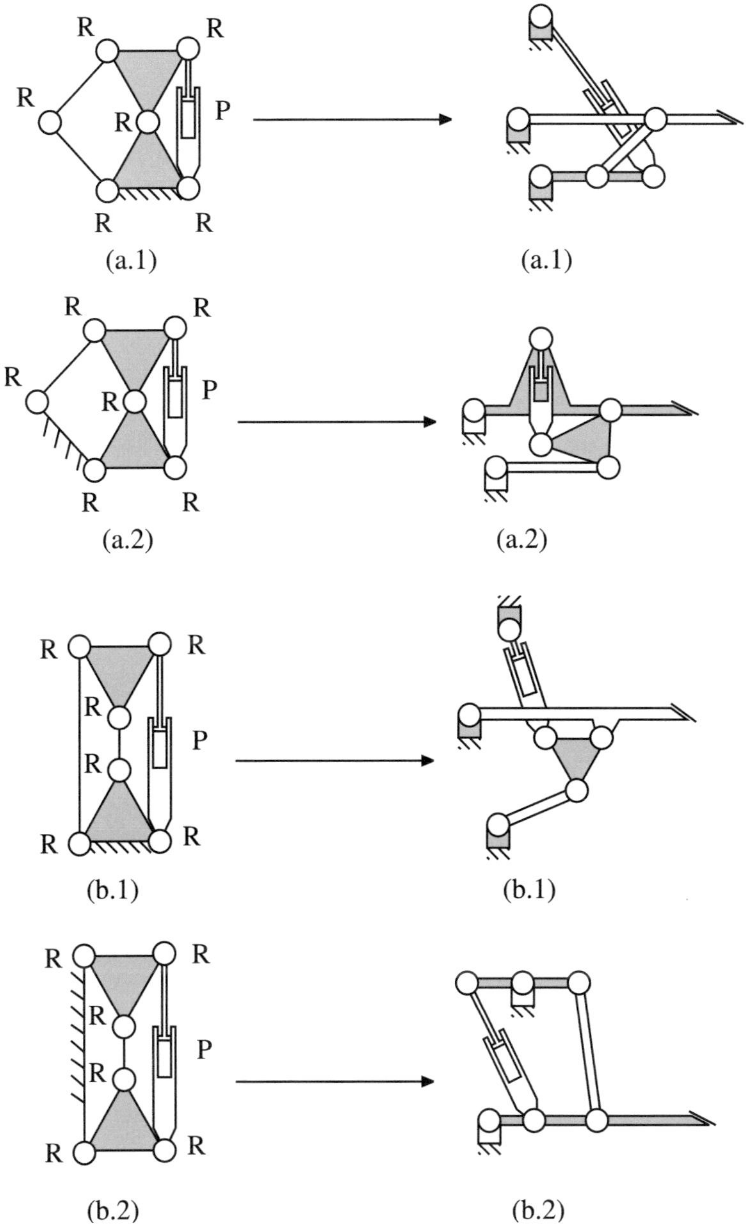

Fig. 6.21 Mechanisms for rear suspension

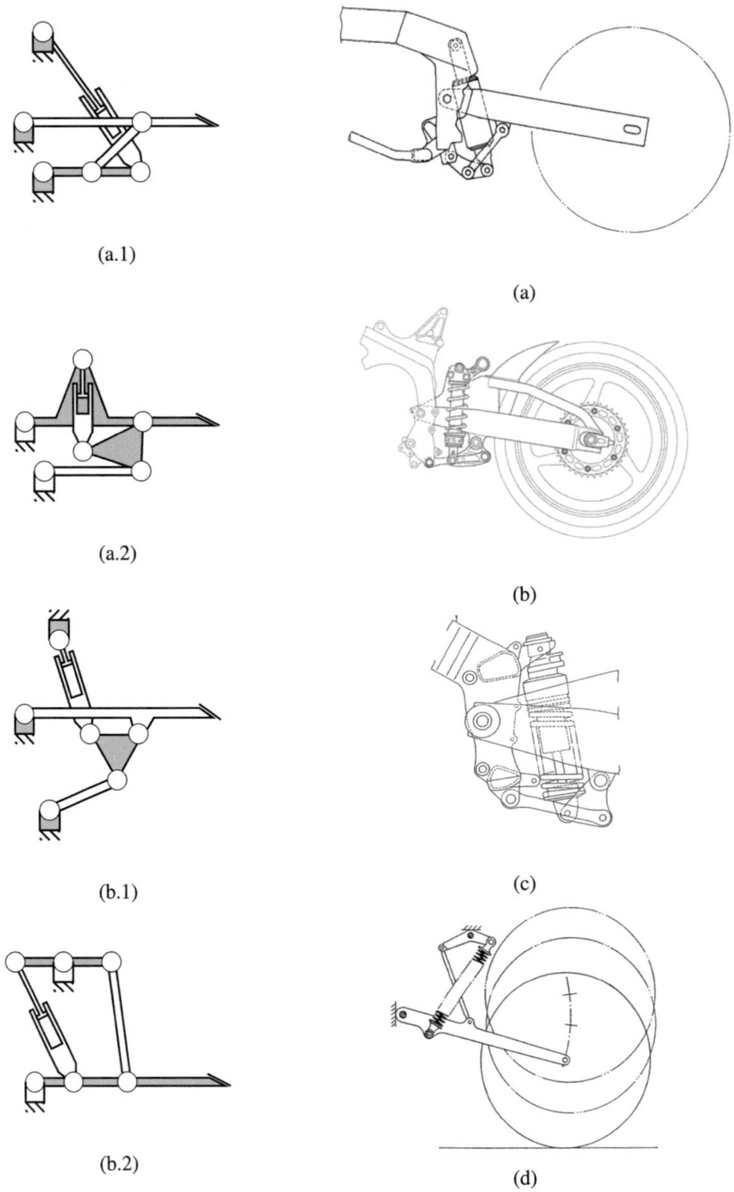

Fig. 6.22 Motorcycle suspensions. **a** Kawasaki Uni-trak. **b** Honda Unit Prolink. **c** Honda Prolink. **d** Suzuki Full Floater

For more information on type synthesis, see [1, 2]. For examples of how to use design requirements to identify and eliminate unfeasible results, see [6–8].

Problems

6.1 Find all proper and non-isomorphic partitions, variations and inversions for the following structural requirements: $M = 3$, $\nu = 2$, $\lambda = 3$ (planar).

6.2 Find, if possible, the proper and non-isomorphic partitions, variations and inversions for the following structural requirements: $M = 3$, $n = 7$, $\lambda = 3$ (planar).

6.3 Find all proper and non-isomorphic partitions, variations and inversions for the following structural requirements: $M = 4$, $\nu = 2$, $\lambda = 3$ (planar).

6.4 For exercise 1, do the type synthesis.

References

1. Hartenberg, R.S., Denavit, J.: Kinematic Synthesis of Linkages. McGraw-Hill, New York (1964)
2. Yan, H.S.: Creative Design of Mechanical Devices. Springer, Singapore (1999)
3. Tsai, L.W.: Mechanism Design: Enumeration of Kinematic Structures According to Function. CRC Press, Boca Raton (2000)
4. Simoni, R., Carboni, A.P., Martins, D.: Enumeration of kinematic chains and mechanisms review. In: Proceedings: 13rd World Congress in Mechanism and Machine Science. IFToMM, Guanajuato (2011)
5. Simoni, R.: Síntese estrutural de cadeias cinemáticas e mecanismos. Master's dissertation. Federal University of Santa Catarina, Florianópolis (2008)
6. Murai, E. H.: Number Synthesis Methods for Mechanism Design: An Alternative Approach. PhD thesis. Federal University of Santa Catarina, Florianópolis (2019)
7. Tischler, C., Samuel, A., Hunt, K.: Selecting multi-freedom multi-loop kinematic chains to suit a given task. Mech. Mach. Theory **36**(8), 925–938 (2001)
8. Murai, E.H.: Projeto de mecanismos de costura com acesso unilateral usando síntese do número e do tipo. Master's dissertation. Federal University of Santa Catarina, Florianópolis (2013)

Chapter 7
Dimensional Synthesis of a Four-Bar Linkage for Two Finitely Separated Positions (2 FSP)

Abstract This chapter addresses the dimensional synthesis of four-bar linkages to achieve motion through two finitely separated positions (2 FSP). It begins by classifying synthesis types—function, path, trajectory, and motion generation—before focusing on the motion generation case. Geometric principles, including rotation poles and mediatrices, are applied to identify the fixed and moving pivots required for the desired motion. Methods for specifying either moving or fixed pivots are described, giving the designer flexibility depending on the constraints of the application. Practical examples such as trapdoor mechanisms are used to illustrate the synthesis steps, supported by figures and coordinate data. Special attention is paid to potential design defects like circuit defect and transmission angle limitations. The chapter emphasizes the iterative and graphical nature of the synthesis process, often supported by software tools like GeoGebra. Overall, it equips readers with practical strategies for two-position kinematic design.

Keywords Types of dimensional synthesis · Synthesis for two finitely separated positions · Fixed pivot specification · Moving pivot specification · Filemon's line analysis · Dimensional synthesis defects

In this chapter we look at concepts covering dimensional synthesis and also the dimensional synthesis of a four-bar linkage specifying two finitely separated positions (2 FSP).

7.1 Concepts About Dimensional Synthesis

In this section we look at general concepts about dimensional synthesis. These concepts are valid for 2, 3 and 4 FSP synthesis.

D. Martins and E. H. Murai, *Mechanisms–Synthesis and Analysis with Applications in Robotics*, Mechanisms and Machine Science 183,
https://doi.org/10.1007/978-3-031-95187-9_7

7.1.1 Types of Dimensional Synthesis

Dimensional synthesis can be classified into four types [1–3]: function generation, path generation, trajectory generation and motion generation.

Function generation
The aim of this synthesis is to determine the mechanism dimensions so that its behavior describes a given function. Thus, for an input angle value, the output bar should return a desired value for the output angle, as a mathematical function. An example is shown in Fig. 7.1,[1] in which a four-bar mechanism is used to calculate the function $\log_{10} x$. The dimensions of the four-bar mechanism are $\overline{O_A O_B} = 1\, l.u.$, $\overline{O_A A} = 3.352\, l.u.$, $\overline{O_B B} = 3.486\, l.u.$ and $\overline{AB} = 0.846\, l.u.$.

Another example is the device used to determine the height and time of high and low tides. Figure 7.2 shows the analog computer used to predict tides.

Coordination of coupler and crank links
The coordination of coupler and crank links is a variation of the function generation, where the output is the coupler link angle instead of the output link angle. This variation is described in detail by [5].

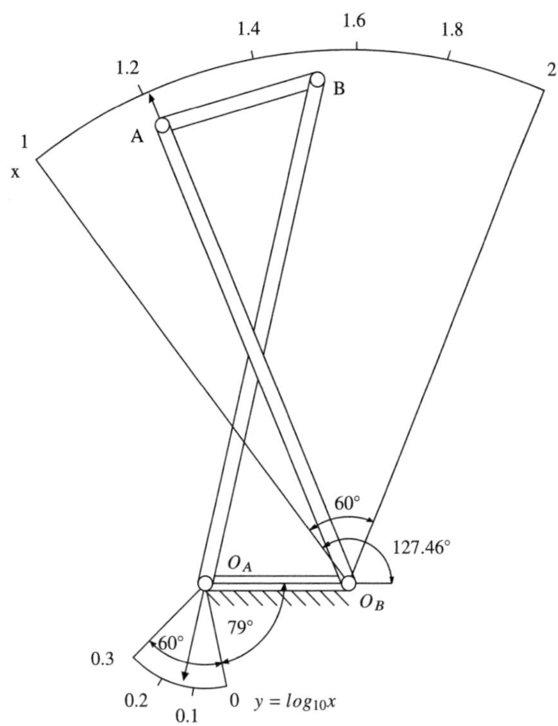

Fig. 7.1 Mechanism for generating the logarithmic function in base 10 of x

[1] Inspired by [4].

Fig. 7.2 Analog computer for tide prediction

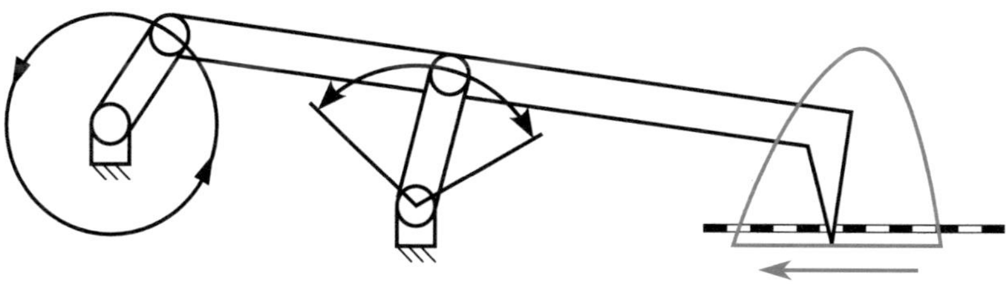

Fig. 7.3 Mechanism for advancing film in camcorders

Path generation

The aim of this synthesis is to determine the dimensions of the mechanism so that a point of interest travels along a specific path when the mechanism is operating. This type of synthesis specifies the path that the point should take, but it does not specify the link orientation along the path. An example is the mechanism for advancing camcorder film, shown in Fig. 7.3. In this mechanism, it is desired that the coupler point travels along a certain path in order to move the film.

Path generation with prescribed timing

Similar to path generation, but with the additional specification of the time on the path or the input link position as the point moves along the path [1]. Such a specification would be useful to make a mechanism that moves a welding torch. In this case, it is not only important to travel along the correct path, but also to have a given speed.

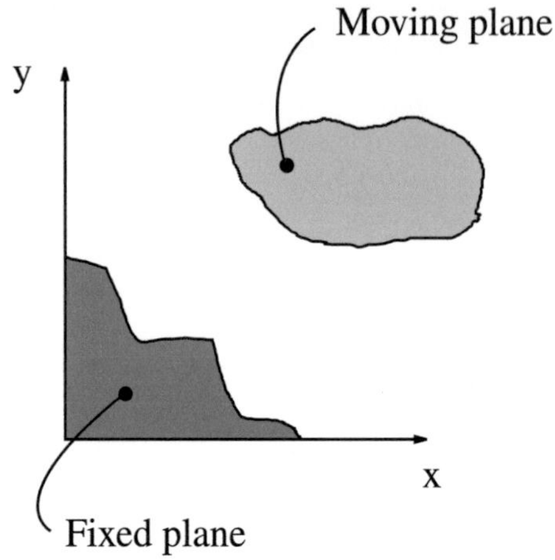

Fig. 7.4 Fixed and moving planes

Motion generation or rigid body guidance
The aim of this synthesis is to determine the dimensions of a mechanism given a set of homologous positions of a plane [3]. In this case, position and orientation are controlled, but only for a finite number of positions. This type of synthesis is used when it is required that the mechanism passes through the given positions with a given orientation and there is no need to control the path between the positions or the time between them.

This books approaches the motion generation dimensional synthesis.

7.1.2 Motion Generation Dimensional Synthesis

The motion generation dimensional synthesis determines the dimensions of a mechanism so that it passes through certain desired positions. The fixed link and the link that must pass through the positions are represented by planes.

Fixed plane: plane containing the fixed reference point. This is the plane by which the mechanism will be mounted.

Moving plane: coupler bar plane. A point or line contained in this plane must do the desired motion.

Both planes can be seen in Fig. 7.4. The dimensional synthesis must determine the dimensions of the four bars for the moving plane to do the desired motion.

Homologous positions are the positions that the moving plane assumes in relation to the fixed plane when the mechanism is operating. Figure 7.5 shows the moving plane assuming three different positions. These three positions are said to be homologous, as they are positions that the same plane assumes when the mechanism is actuated.

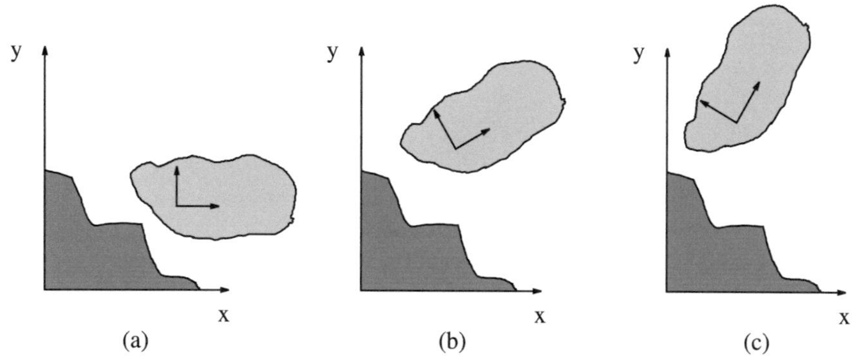

Fig. 7.5 Homologous positions of the moving plane. **a** position 1. **b** position 2. **c** position 3

Homologous positions can refer not only to planes, but also to points on the plane, called **homologous points**. For motion generation, the dimensional synthesis problem can be specified in two ways:

Homologous line segments: uses homologous line segments, defining both the position and the desired orientation via the two endpoints of the line segment. An example is seen in Fig. 7.6, in which the moving plane must pass through positions 1, 2 and 3, which are defined by the line segments $\overline{C_1D_1}$, $\overline{C_2D_2}$ and $\overline{C_3D_3}$.

Homologous points and angles: uses a point to specify the desired position and an angle to specify the desired orientation. An example can be seen in Fig. 7.7.

Once the problem has been specified, the geometric approach or the algebraic approach can be used to solve it. For the synthesis of 2 FSP and 3 FSP, the geometric approach will be used. This approach consists of determining the positions of the revolute pairs (also called **pivots**). Determining the pivots position indirectly determines the links dimensions.

The pivots of the four-bar mechanism can be classified into fixed and moving pivots.

Fixed pivots are the pivots that are in the fixed plane (fixed bar), i.e. these pivots will always have the same pair of coordinates (x, y) in time. The fixed pivot that connects the fixed bar and the input bar is called the input bar fixed pivot, and it is represented by the symbol A_0. The fixed pivot that connects the fixed bar and the output bar is called the output bar fixed pivot and it is represented by the symbol B_0.

Moving pivots are the pivots that are on the moving plane (coupler bar), so these pivots change their position as the mechanism is driven. The moving pivot that connects the input and coupler bars is called the input bar moving pivot, and it is represented by the symbols A_1, A_2 and A_3 for each of the three positions. The moving pivot that connects the output and coupler bars is called the output bar moving pivot, and it is represented by the symbols B_1, B_2 and B_3 for each of the three positions. Often A_m and B_m are used to represent the moving pivots in an intermediate position between the desired positions.

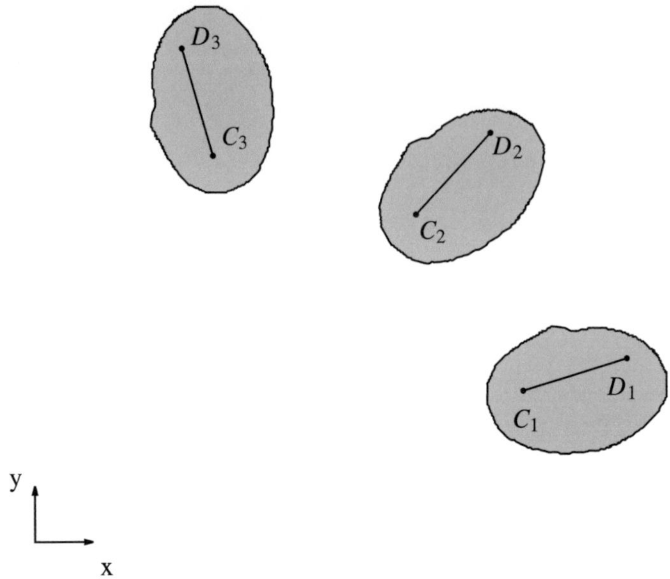

Fig. 7.6 Specification using homologous line segments

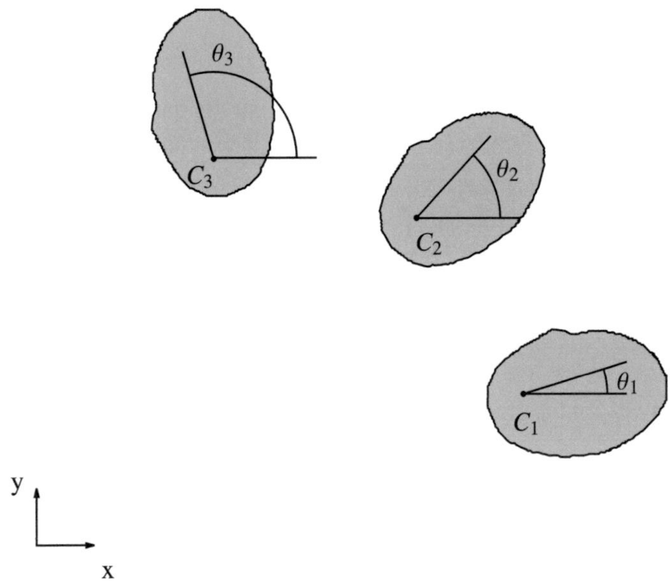

Fig. 7.7 Specification using homologous points and angles

7.1.3 Mechanism Selection

When designing a new machine, it is recommended to follow a design methodology. Methodologies usually have four initial stages: informational, conceptual, preliminary and detailed design.[2] The number and type syntheses, which we looked at in Chap. 6, is related to the conceptual and preliminary design stages. Dimensional synthesis is related to the detailed design stage. It is therefore assumed that the designer has carried out the informational, conceptual and preliminary design stages and has concluded that the mechanism to be used is the four-bar mechanism. Therefore, the dimensional synthesis presented in this chapter and in Chaps. 8 and 9 is for the four-bar mechanism.

For mechanisms other than the four-bar mechanisms, dimensional synthesis becomes more complicated and optimization methods must be used to obtain the mechanism whose motion is the closest to the desired motion. The reader can find out more about this topic at [3]. Therefore, this book approaches only the dimensional synthesis for the four-bar mechanism, since this mechanism has an analytical and geometric solution that is easy to obtain. Although it may seem limiting to obtain the dimensional synthesis only for the four-bar mechanism, the reader should remember that the four-bar mechanism is capable of traversing curves of up to sixth order and that the four-bar mechanism or the combination of two four-bar mechanisms is widely used (either Watt or Stephenson).

Finally, the designer should bear in mind that the procedures presented for dimensional synthesis serve to provide an approximation of the desired dimensions. Although the procedure provides an exact solution, it often results in links whose sizes are not whole numbers. It is recommended to use integer dimensions, such as 240 mm instead of 238,57 mm, as they facilitate design, manufacture, inspection and maintenance. Thus, the procedures serve as guides for the mechanism dimensional design, and the designer should be aware that rounding off the dimensions obtained results in small positioning errors. It is up to the designer to assess whether these errors are relevant or they can be disregarded for the project in question.

7.2 Synthesis for Two Finitely Separated Positions

The dimensional synthesis of four-bar linkages for two finitely separated positions, or *synthesis 2 FSP*, aims to determine the dimensions of a four-bar mechanism so that the coupler link passes through the two desired positions.

Sections 7.2.1 and 7.2.3 show the procedures for determining the dimensions of the four-bar mechanism by specifying the moving pivots. Section 7.2.4 shows the procedure for determining the four-bar mechanism dimensions by specifying the fixed pivots. When designing the four-bar mechanism, it is not necessary to do

[2] There are other subsequent stages, as can be seen in [6].

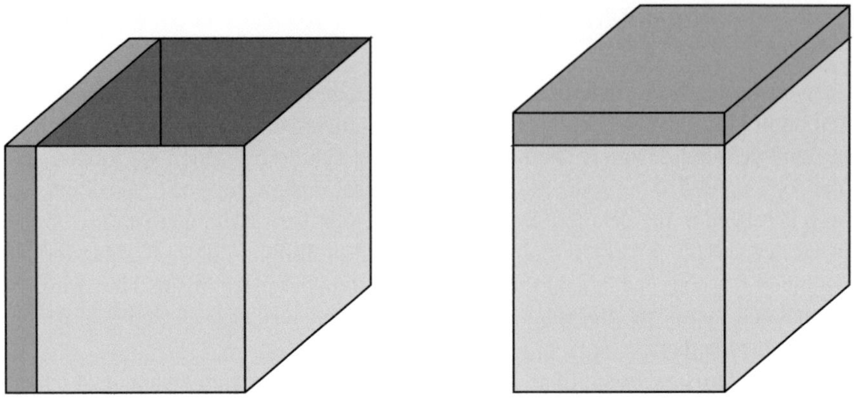
(a) (b)

Fig. 7.8 Closing the box lid problem

both procedures (specifying the moving and fixed pivots). Therefore, one or the other procedure is followed, depending on whether one needs more freedom when specifying the moving or fixed pivots.

7.2.1 Specifying Moving Pivots

Consider the problem of guiding the lid of a box by removing the lid from the side of the box and placing it at the top opening, as indicated in Fig. 7.8.

Initially, the problem is written according to the design specifications, as presented in Sect. 7.1.2. In this case, one can represent the two desired positions with a pair of points in each position, with C_1 and D_1 in position 1 and C_2 and D_2 in position 2, as shown in Fig. 7.9.

Any pair of homologous points can be used to represent the position, including points that are not on the box lid. In the latter case, it would be enough to extend the coupler bar so that the moving plane also covers the lid.

The synthesis procedure is carried out using GeoGebra. The points C_1, D_1, C_2 and D_2 are inserted, corresponding to the desired positions, as shown in Fig. 7.10.

Next, the **pole of rotation** is determined. Every motion in space can be represented instantaneously by a rotation around an axis. In the planar workspace, a motion can be represented instantaneously by a rotation around a point, the pole of rotation. This representation is said to be instantaneous because the pole of rotation usually moves in the plane. If the motion is a translation, it can be represented by a rotation around a point at infinity.

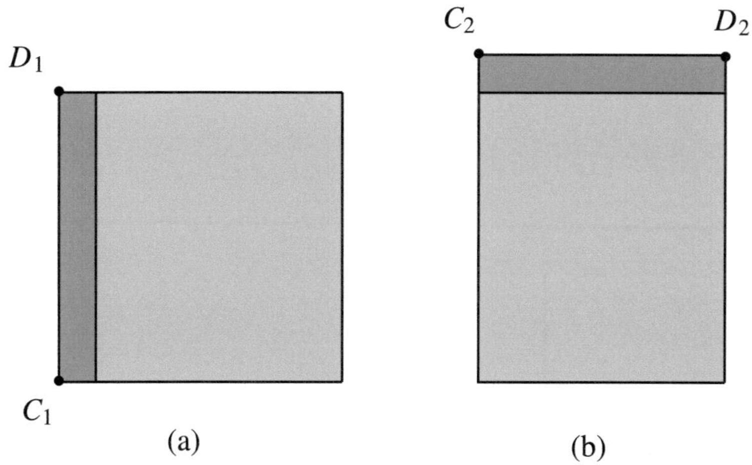

Fig. 7.9 Representation of desired positions

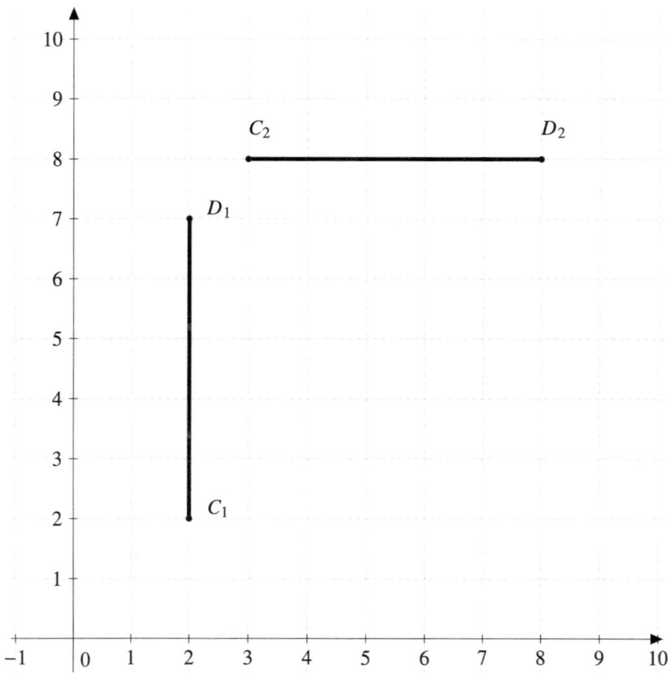

Fig. 7.10 Desired positions entered into GeoGebra

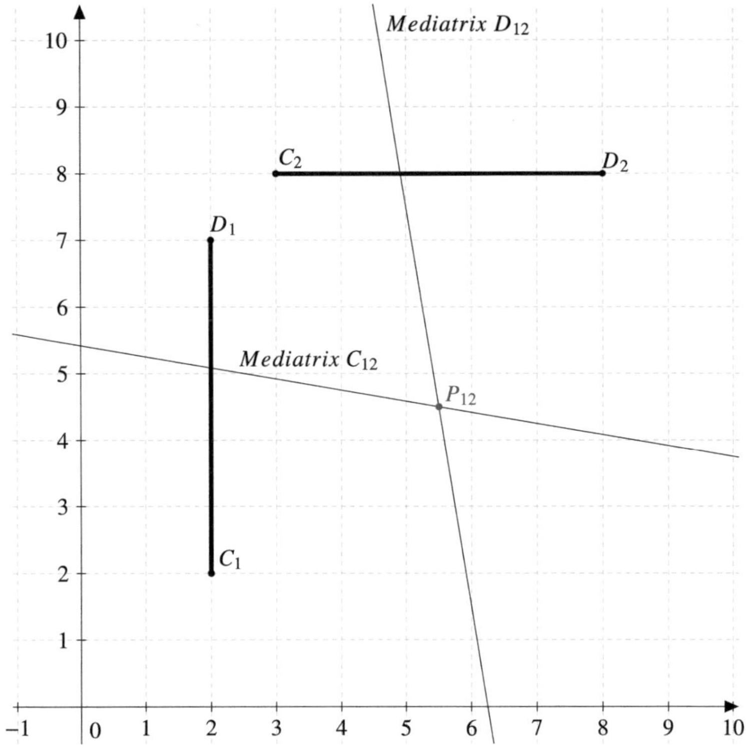

Fig. 7.11 Mediatrices

To determine the pole of rotation, one has to draw the mediatrix of points C_1 and C_2 and then the mediatrix of points D_1 and D_2.[3] The two mediatrices are shown in Fig. 7.11. A common mistake is to make the mediatrices using C_1 and D_1 or C_2 and D_2.

The meeting of the two mediatrices is the pole of rotation. Starting from position 1, you can reach position 2 by rotating around pole P_{12}. The magnitude of this rotation, θ_{12}, can be measured by the angle between the line segments $\overline{P_{12}C_1}$ and $\overline{P_{12}C_2}$. The same angle can also be measured between the line segments $\overline{P_{12}D_1}$ and $\overline{P_{12}D_1}$, in fact any homologous point can be used to measure the rotation θ_{12} around P_{12}. Figure 7.12 shows the angle of rotation measured in the two ways mentioned above, using points C and D.

Next, the position of the B_1 pivot is chosen. This pivot can be positioned anywhere that is convenient. Returning to the concept of the pole of rotation, if the moving plane rotates θ_{12} around P_{12} to leave position 1 and go to position 2, all points on the

[3] Mediatrix is the locus of the points that are the same distance from two different points A and B (in the case of the synthesis C_1 and C_2 or D_1 and D_2). One way to determine the mediatrix is to draw a line segment between C_1 and C_2 and draw a line intersecting the previous line segment in the middle of it at 90°. In GeoGebra, simply use the "Mediatrix" tool and select the two points.

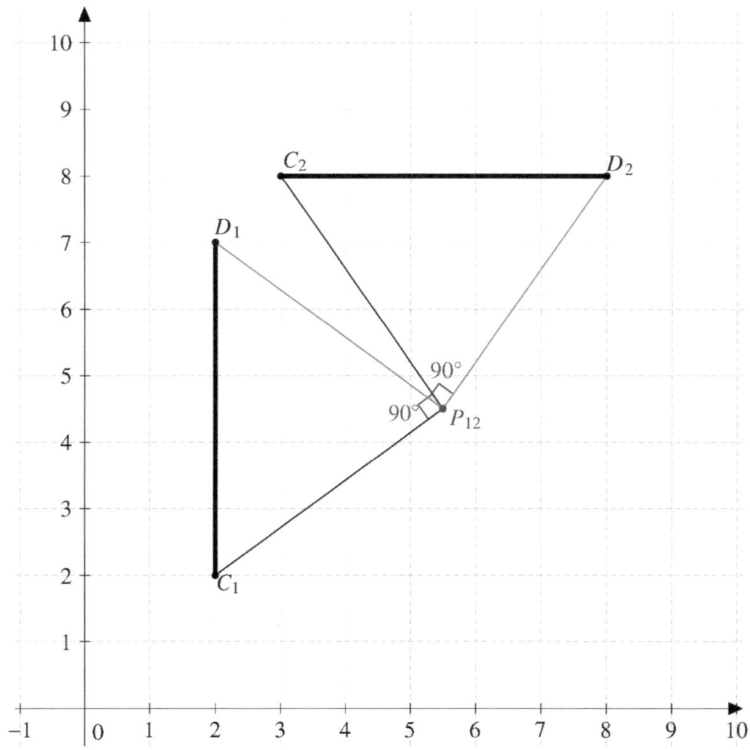

Fig. 7.12 Angle of rotation θ_{12} around pole P_{12}

moving plane will also do this same motion (as seen for C and D). So, if we choose the position of point B_1, to get the position of point B_2 we just need to rotate B_1 from θ_{12} around P_{12}. This procedure is shown in Fig. 7.13. Note that the position of B_2 relative to D_2 is similar to the position of B_1 relative to D_1. If these positions are not visually similar, a mistake has been made and the procedure should be repeated.

Once we have the position of B_2, we draw the mediatrix between B_1 and B_2. This mediatrix must pass through the pole of rotation, after all, B_1 and B_2 are homologous points just as points C_1 and C_2 or D_1 and D_2 were when the pole of rotation was determined. The pivot B_0 is chosen on this new mediatrix. In position 1, the output bar is made up of pivots B_0 and B_1, i.e. the distance between these two pivots is the length of the output bar. In position 2, the output bar is made up of pivots B_0 and B_2, and the distance between these two pivots is the length of the output bar. As the length of the output bar must remain constant throughout the operation, the distance between B_0 and B_1 must be equal to the distance between B_0 and B_2. As already mentioned, the locus of the points equidistant from B_1 and B_2 is the mediatrix, i.e. B_0 must lie on the mediatrix for the length of the output bar to remain constant over time. Figure 7.14 shows the mediatrix of B_1 and B_2 as well as the chosen B_0.

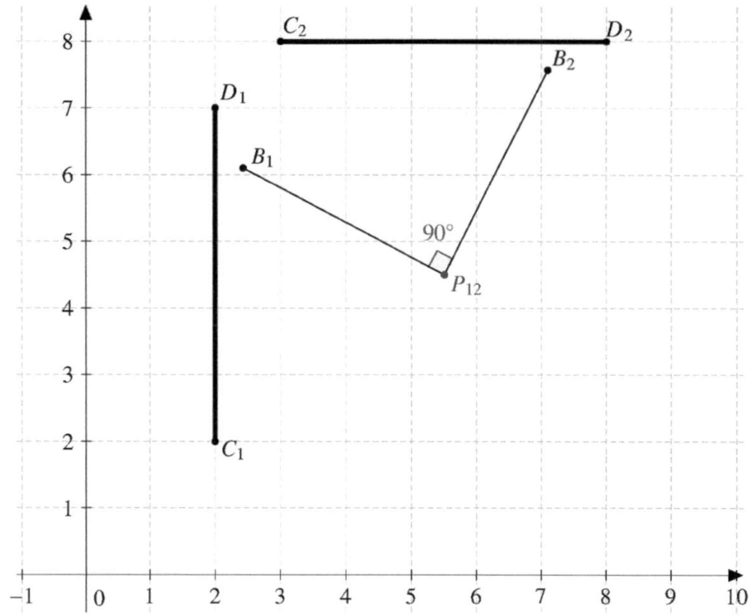

Fig. 7.13 Rotating B_1 around pole P_{12}

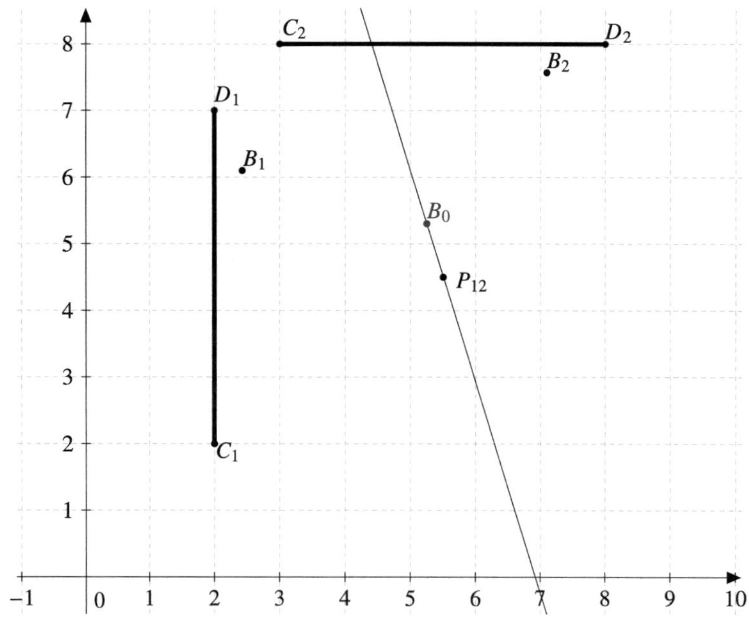

Fig. 7.14 Choosing the pivot B_0

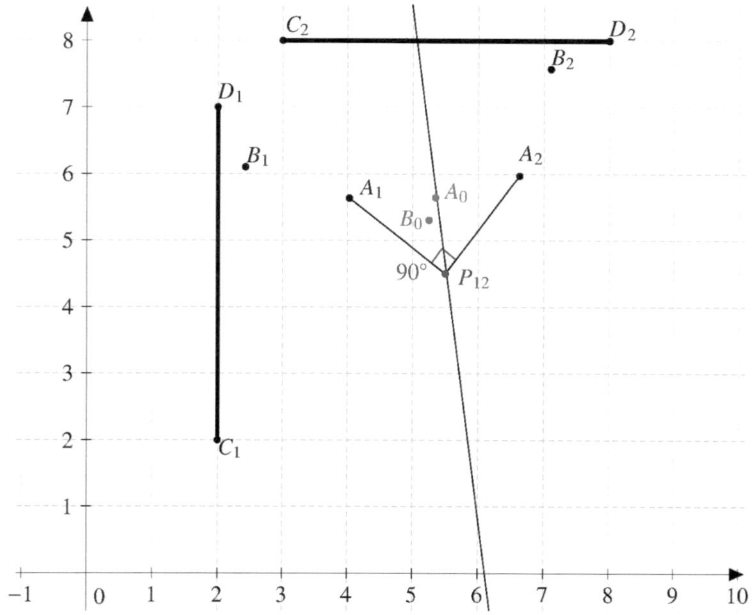

Fig. 7.15 Determining the input bar

Once the output bar has been determined, the procedure is repeated to find the input bar. Thus, A_1 is chosen, θ_{12} is rotated around the pole to find A_2 and, in the mediatrix of A_1A_2, A_0 is chosen, as shown in Fig. 7.15. This determines the dimensions of the four links. The mechanism in position 2 is shown in Fig. 7.16.

In order for the moving plane to pass through C_1, D_1, C_2 and D_2, it is sufficient to extend the moving plane of the coupler bar to these points so that it covers them, as shown in Figs. 7.17 and 7.18.

However, when simulating the mechanism, it is not possible to leave position 1 and reach position 2. To take the mechanism from position 1 to position 2, it is necessary to disconnect the coupler and output links and reconnect them in another working mode.

Figure 7.19 shows a circle centered on B_0 and with a radius equal to the size of the output bar (the circle is the locus of the pivot B_m). The same figure shows a circle centered on A_1 and with a radius equal to the size of the coupler bar. The two intersections of these circles are the possible locations for B_1, resulting in the two possible working modes.

In position 1, the working mode of point B_1 was used, which would correspond to point B_2' in position 2 (see Fig. 7.20), showing the use of two different working modes. However, reassembling the mechanism in the correct working mode during operation is impractical. It can be seen that using B_1' in position 1 and B_2 in position 2 would result in two different moving planes, and deforming the moving plane during the mechanism's operation is also unfeasible. Therefore, this mechanism is not a solution.

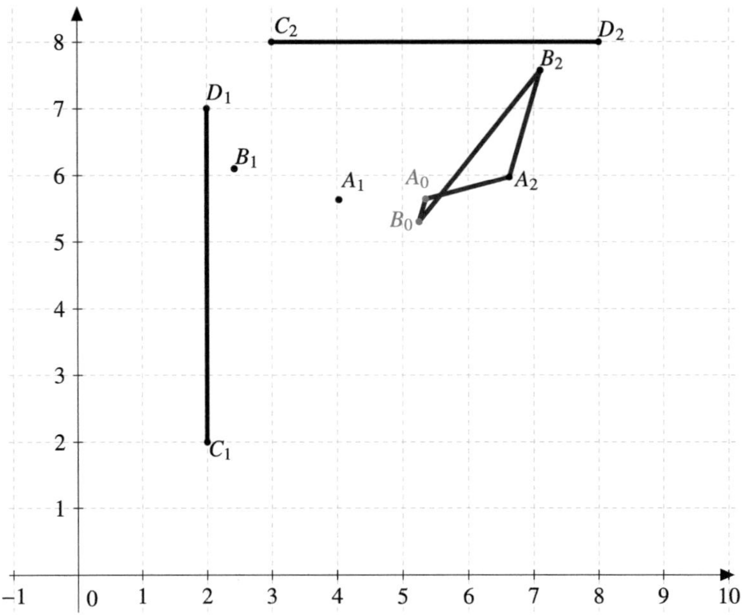

Fig. 7.16 Four-bar mechanism in position 2

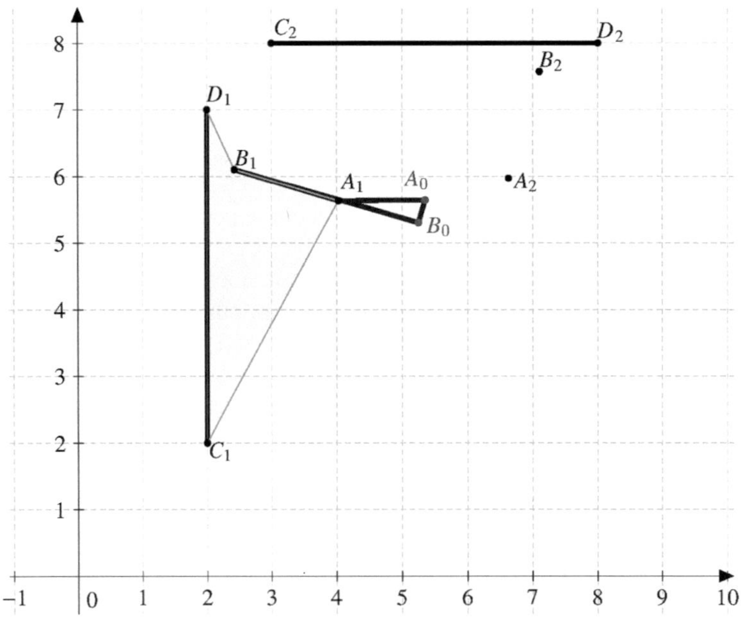

Fig. 7.17 Moving plane reaching position 1

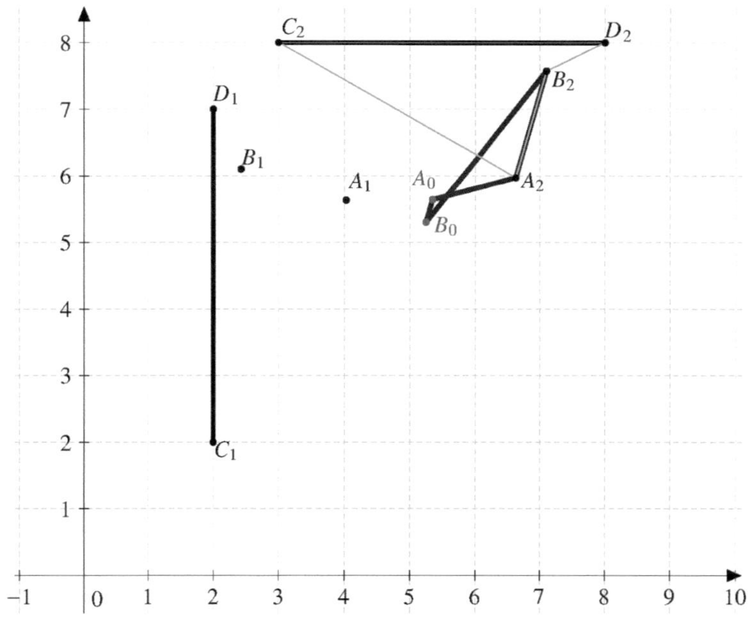

Fig. 7.18 Moving plane reaching position 2

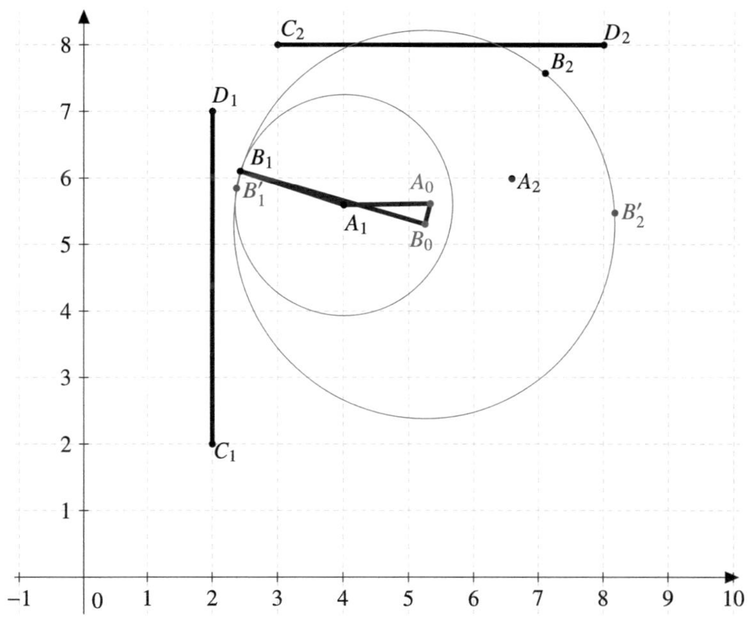

Fig. 7.19 Moving plane reaching position 1

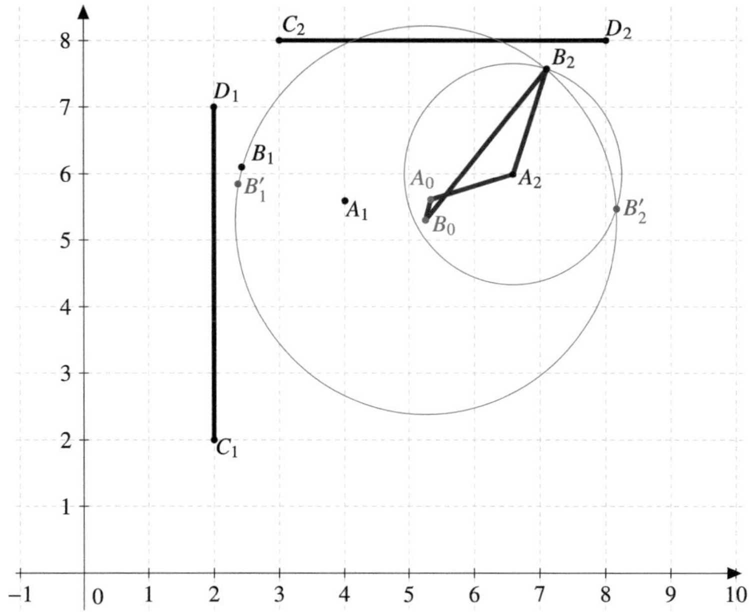

Fig. 7.20 Moving plane reaching position 2

7.2.2 *Filemon's Lines for 2 FSP*

In Sect. 7.2.1, a dimensional synthesis of a four-bar mechanism was made and a defect was noted. This defect is known as a circuit defect, characterized by the fact that the mechanism uses two working modes during its operation, making it unfeasible.

Another defect that can occur in the dimensional synthesis of the four-bar mechanism is the branching defect. This defect is characterized by the alignment of the output and coupler bars. This alignment is undesirable because it is a singularity position, as explained in section 2.4.

The two defects presented (circuit and branch) can be avoided by using the **Filemon's lines** analysis. Filemon's lines, a method developed by Elizabeth Filemon in 1970 [7], aims to determine the region of the plane where the pivot A_1 cannot be positioned. If the pivot is positioned in this region, the mechanism may have one of the defects mentioned.

The principle of the method is to determine the rotation of the coupler bar in relation to the output bar when the mechanism is actuated. Once this rotation has been determined (ψ_{12}), the forbidden region corresponds to $-\psi_{12}$. This concept is exemplified in Fig. 7.21. Note that if A_1 is placed in the highlighted region, when moving ψ_{12} degrees, point A_m crosses the alignment line (0°), making A_m, B_m and B_0 collinear. If placed outside the highlighted region, A_1 does not cross the 0° line.

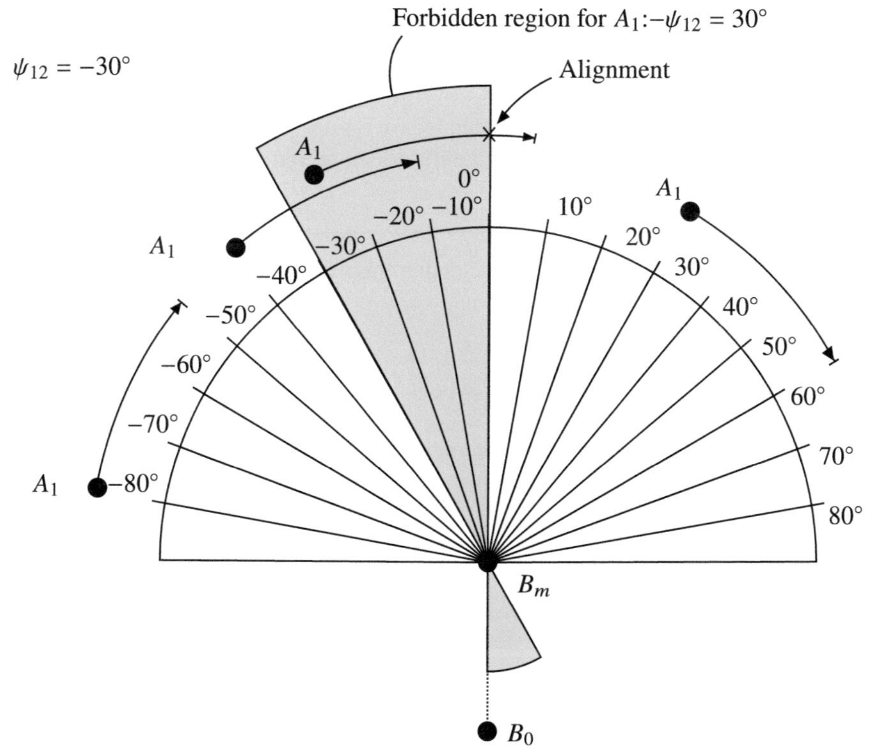

Fig. 7.21 Forbidden region for positioning A_1 according to Filemon's lines

To determine the rotation ψ_{12}, measure the angle between the output bar and the coupler bar at positions 1 and 2 (ψ_1 and ψ_2, respectively). As the position of the pivot A_1 and A_2 is not known, it is not known where the coupler bar is. Therefore, it is used two homologous points at the coupler bar to measure the angles ψ_1 and ψ_2. There are several ways to measure ψ_1 and ψ_2. For didactic purposes and to avoid errors, it is recommended to always follow one way, preferably the step-by-step below:

1. Draw the **ray** that **starts** from B_0 and passes through B_1.
2. Draw the **ray** that **starts** from B_1 and passes through C_1.
3. Measure the angle between the ray $\overrightarrow{B_0B_1}$ and the ray $\overrightarrow{B_1C_1}$ (note that the reference is $\overrightarrow{B_0B_1}$). This is the angle ψ_1.
4. Draw the **ray** that **starts** from B_0 and passes through B_2.
5. Draw the **ray** that **starts** from B_2 and passes through C_2.
6. Measure the angle between the ray $\overrightarrow{B_0B_2}$ and the ray $\overrightarrow{B_2C_2}$ (note that the reference is $\overrightarrow{B_0B_2}$). This is the angle ψ_2.
7. The angle ψ_{12} is given by: $\psi_{12} = \psi_2 - \psi_1$. This is the angle that moves the coupler bar from position 1 to position 2.

8. The forbidden region for A_1 lies between the **line** $\overline{B_0B_1}$ and the **line** that passes through B_1 and has slope $-\psi_{12}$ with respect to the ray $\overrightarrow{B_0B_1}$. Note that the forbidden region is between two **lines**, covering both sides of the circle, as shown in Fig. 7.21. Also, note that the forbidden region is given by rotating ψ_{12} to the opposite side. Thus, if ψ_{12} is negative, the forbidden region will be given by rotating $+\psi_{12}$.

Once the forbidden region for A_1 has been determined, it is enough to position this pivot outside this region to avoid the alignment problem.

General observations:

- If another homologous point were chosen, such as D_1, the values of ψ_1 and ψ_2 would be different, but ψ_{12} would remain the same, resulting in the same forbidden region.
- If A_1 is chosen within the forbidden region, when the mechanism is actuated the moving pivot A_m will rotate, leaving A_1 and heading towards A_2. But at some point during this rotation, A_m will be on the extension of $\overline{B_0B_m}$, i.e. the links will be aligned.
- If it is desired to determine the forbidden region for the pivot A_2, just repeat the last two steps for position 2, i.e. $\psi_{21} = \psi_1 - \psi_2$, and the forbidden region is given between the line $\overline{B_0B_2}$ and the line that passes through B_2 and has slope $-\psi_{21}$ with respect to the ray $\overrightarrow{B_0B_2}$.

Typical mistakes made when doing Filemon's lines analysis:

- To use as a reference the ray that starts from the pole of rotation and passes through B_1 instead of using the ray that starts from B_0 and passes through B_1. Making the same mistake for B_2 is also common.
- To wrongly measure angles ψ_1 and ψ_2, one clockwise and the other counterclockwise, because they were visually the smallest angles.
- To use lines instead of rays to measure the angles, which can lead to measuring an angle plus 180° or $180° - \psi_1$ or, in the case of position 2, $180° - \psi_2$.
- Use the ray of the coupler as a reference and not the ray of the output bar.

7.2.3 Specifying Moving Pivots with Filemon's Lines Analysis

Filemon's lines analysis is carried out in this section for the mechanism of Sect. 7.2.1 with the given output bar. Filemon's lines analysis aims to identify the forbidden region for inserting A_1. To create the Filemon's lines, one can use the homologous points C_1 and C_2, as mentioned in Sect. 7.2.2. The angles ψ_1 and ψ_2 are then determined, as shown in Fig. 7.22.

This gives $\psi_1 = 99.73°$ and $\psi_2 = 123.21°$, which results in $\psi_{12} = \psi_2 - \psi_1 = 23.48°$. Therefore, the forbidden region is between the line that contains the output

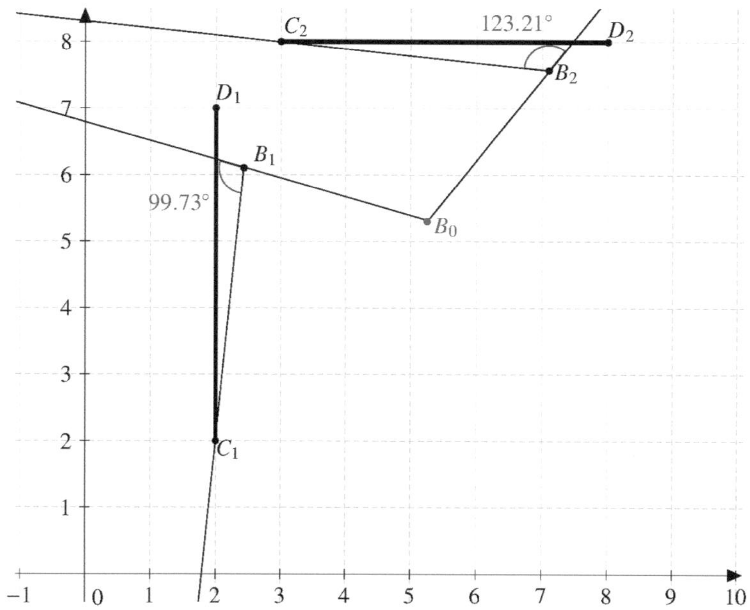

Fig. 7.22 Determining ψ_1 and ψ_2

bar and the line that passes through B_1 and has a slope of $-23.48°$ (i.e. clockwise), as shown in the highlighted region in Fig. 7.23.[4]

The following procedure is similar to the one used for the output bar. The position for the pivot A_1 is chosen outside the forbidden region. A_1 is the input link moving pivot, i.e. it is contained in the moving plane. Therefore, just like C_1, D_1 and B_1, point A_1 will rotate around the pole. The point A_2 is determined by rotating θ_{12} around P_{12}, the result is shown in Fig. 7.24.

Finally, draw the mediatrix of points A_1 and A_2. Similar to the case of B_1 and B_2, this new mediatrix must pass through the pole and the pivot A_0 must be chosen on it, see Fig. 7.25. Once A_0 is chosen, the input bar is determined. If one has all four pivots, A_0, A_1, B_0 and B_1, one has a four-bar mechanism, as shown in Fig. 7.26.

In order for the moving plane to pass through C_1, D_1, C_2 and D_2, it is sufficient to extend the moving plane of the coupler bar to these points so that it covers them, as shown in Figs. 7.27 and 7.28.

It can be seen that three degrees of freedom are used to determine the output bar. Two of them are in the positioning of the moving pivot, B_1, because when positioning this pivot one chooses the coordinates x and y (choice of a point on the plane). The third degree of freedom lies in the choice of the fixed pivot, B_0, because this pivot must lie on the mediatrix of B_1 and B_2 (choice of a point on a line). The same goes for the input bar.

[4] To hatch regions in GeoGebra, one can use the command *integralBetween*, in the command bar (bottom corner of the screen).

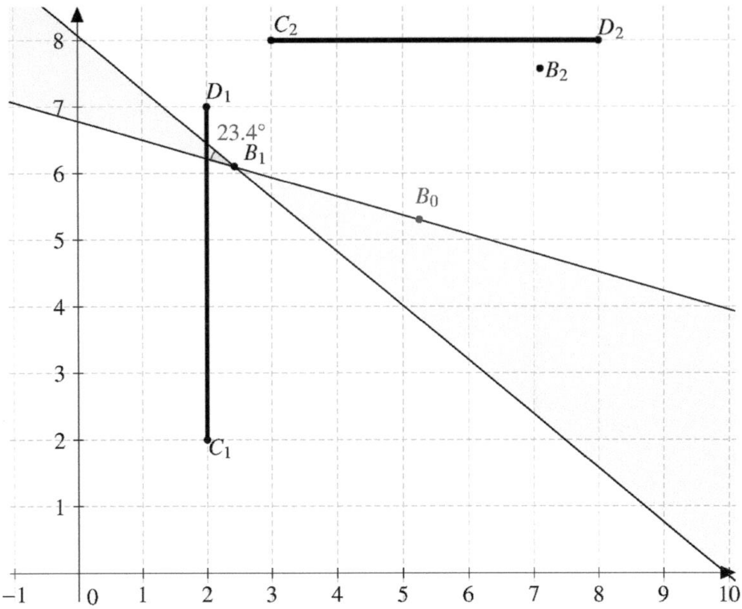

Fig. 7.23 Filemon's lines analysis

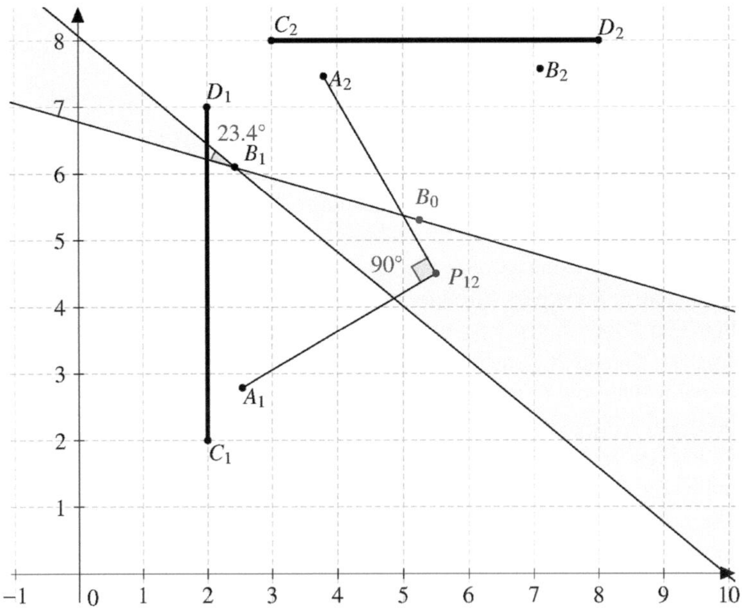

Fig. 7.24 Determination of A_1 and A_2

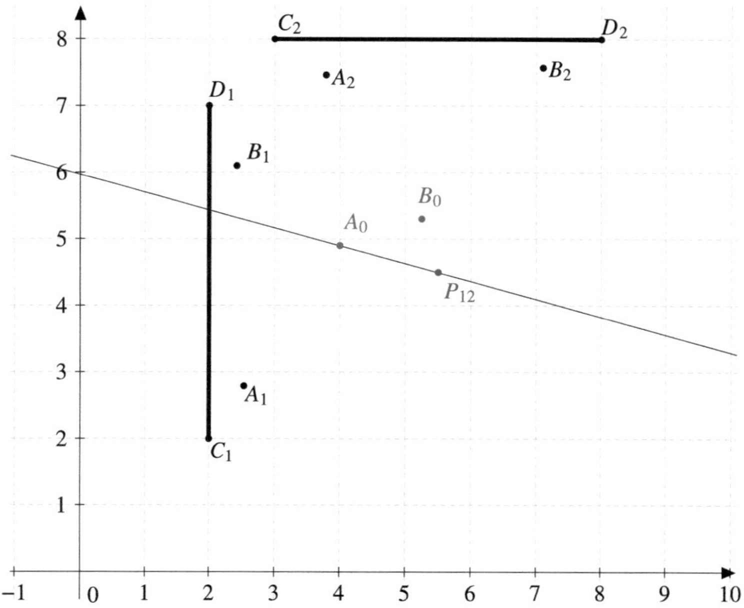

Fig. 7.25 Determination of A_0

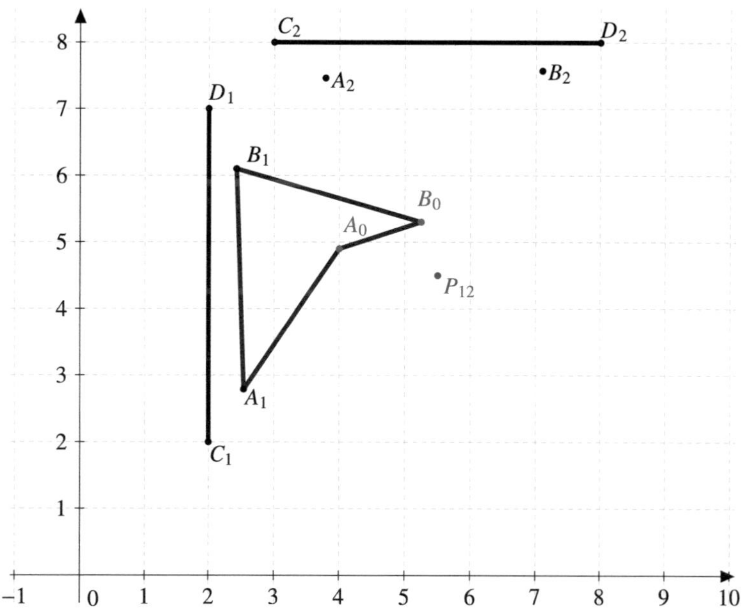

Fig. 7.26 Four-bar mechanism

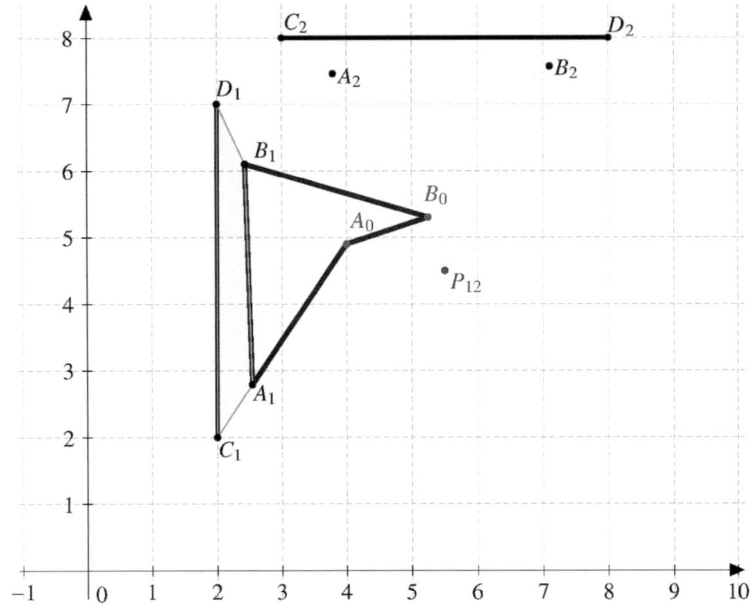

Fig. 7.27 Moving plane reaching position 1

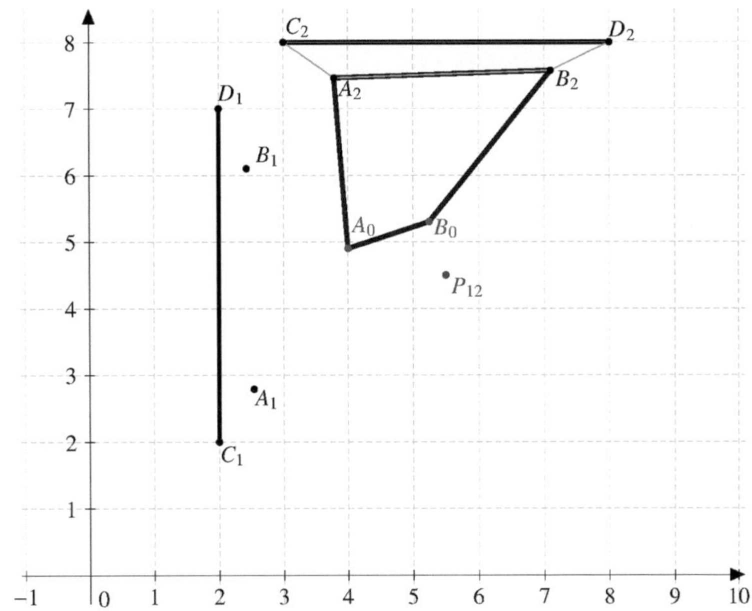

Fig. 7.28 Moving plane reaching position 2

In the next chapter, we want the moving plane to pass through 3 positions. With each position, more constraints are inserted into the project. This reduces the number of choices made. This implies less freedom in the design and, when something unforeseen occurs, it is more difficult to find a solution (such unforeseen events will be discussed in Sect. 7.3).

It is possible to replace the four-bar mechanism with a single pivot at the pole of rotation, so the mechanism would have two links, the base and the moving plane. However, as stated in Sect. 7.1.3, it is assumed that one wants to use the four-bar mechanism. Therefore, the single-pivot solution is discarded here.

The following is a summary of the two-position synthesis specifying the moving pivots.

1. Determine the pole P_{12}.
2. Determine the angle θ_{12}.
3. Choose a position for B_1.
4. Determine B_2 from θ_{12} and P_{12}.
5. Draw the mediatrix of B_1 and B_2 and on it choose B_0.
6. Determine the forbidden region for A_1 by analyzing Filemon's lines.
7. Choose A_1 outside the forbidden region.
8. Determine A_2 from θ_{12} and P_{12}.
9. Draw the mediatrix of A_1 and A_2 and on it choose A_0.

For more information on the synthesis of two finitely separated positions, see [1, 2, 4]. The reader can also look up the works of Ludwig Burmester,[5] the forerunner in studies of the synthesis of motion generation in the plane using geometry [2].

For more information about GeoGebra, visit the website https://www.geogebra.org/.

7.2.4 Specifying Fixed Pivots

The 2 FSP synthesis presented in Sect. 7.2.3 is used when one needs more freedom to specify the position of B_1 and A_1 moving pivots. Next, the same box problem is solved by specifying the fixed pivots more freely.

Returning to the problem of designing a mechanism for closing the box lid, one must first determine the rotation pole and the rotation angle θ_{12}, as explained in Sect. 7.2.3. Next, the position for the fixed pivot of the output link is chosen, as shown in Fig. 7.29.

A line is drawn through B_0 and the pole (this line is equivalent to the mediatrix of B_1 and B_2 when the moving pivots are specified). Two rays are drawn from P_{12},

---

[5] German mathematician renowned for his studies in the field of geometry and kinematics, developed Burmester's theory which uses geometry techniques for the synthesis of articulated mechanisms [8].

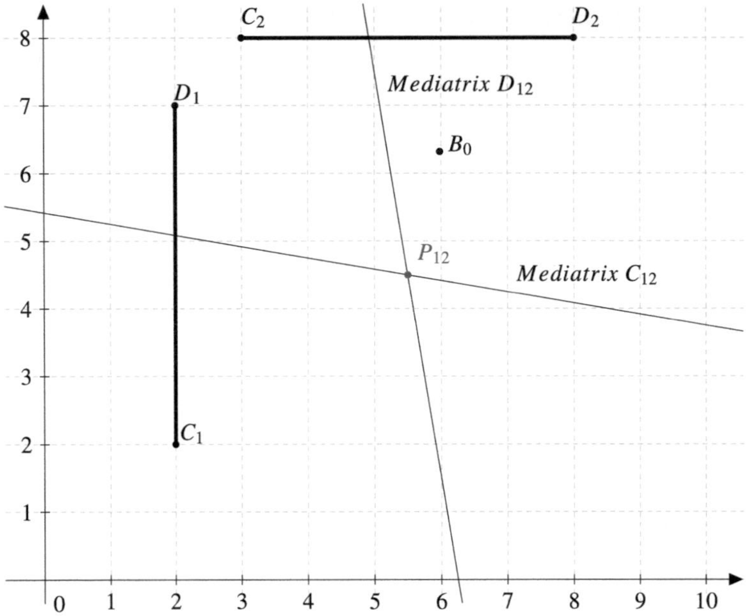

Fig. 7.29 Choosing the fixed pivot position

one at $+\theta_{12}/2$ and the other at $-\theta_{12}/2$, both in relation to the line $\overline{P_{12}B_0}$, as shown in Fig. 7.30. As in this example $\theta_{12} = 90°$, the rays drawn are $\pm 45°$ from $\overline{P_{12}B_0}$.

The output bar will rotate around B_0 and will always be the same length. Therefore, the pivot B_m will always be on a circle centered on B_0 and whose radius is the size of the output bar. So a circle is drawn with a chosen radius and centered on B_0. The two intersections of this circle with the two rays previously drawn will be the points B_1 and B_2. Note that this geometric construction has been done so that the mediatrix of B_1 and B_2 passes through B_0 and the pole. The circle and the points B_1 and B_2 are shown in Fig. 7.31. This procedure determines the output bar.

Repeating the same procedure for the input bar, choose A_0 and draw the line $\overline{P_{12}A_0}$ and the two rays starting from the pole, $\pm 45°$ in relation to the line $\overline{P_{12}A_0}$, as shown in Fig. 7.32.

Next, make a circle centered on A_0 and with the desired radius. The intersection of this circle with the two rays will determine the positions of A_1 and A_2, as shown in Fig. 7.33.

Filemon's lines must be analyzed to check that the pivot A_1 is outside the forbidden region, as shown in Fig. 7.34. Finally, we have the mechanism shown in Fig. 7.35.

Note that it is possible to determine the output bar by specifying the moving pivot and then determine the input bar by specifying the fixed pivot. The opposite is also possible. Thus, by combining these two methods for the input and output bar, the 2 FSP synthesis can be solved in four different ways.

The following is a summary of 2 FSP synthesis by specifying fixed pivots.

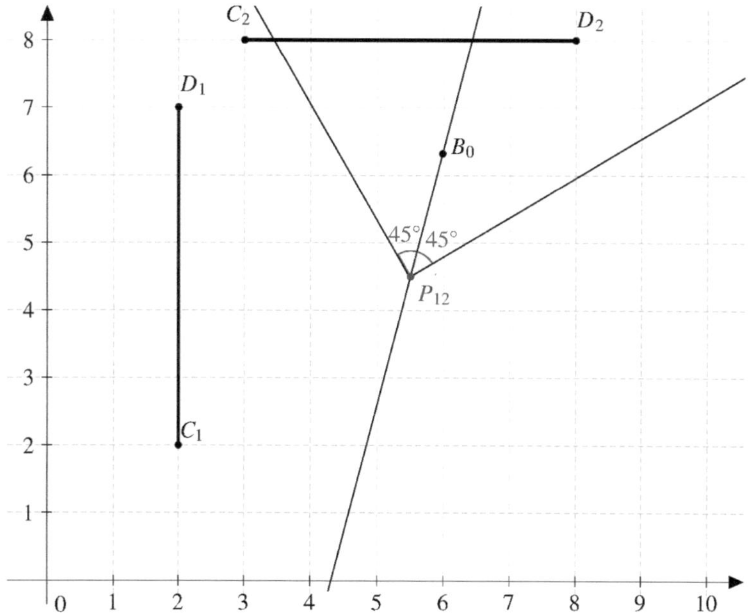

Fig. 7.30 Drawing auxiliary lines

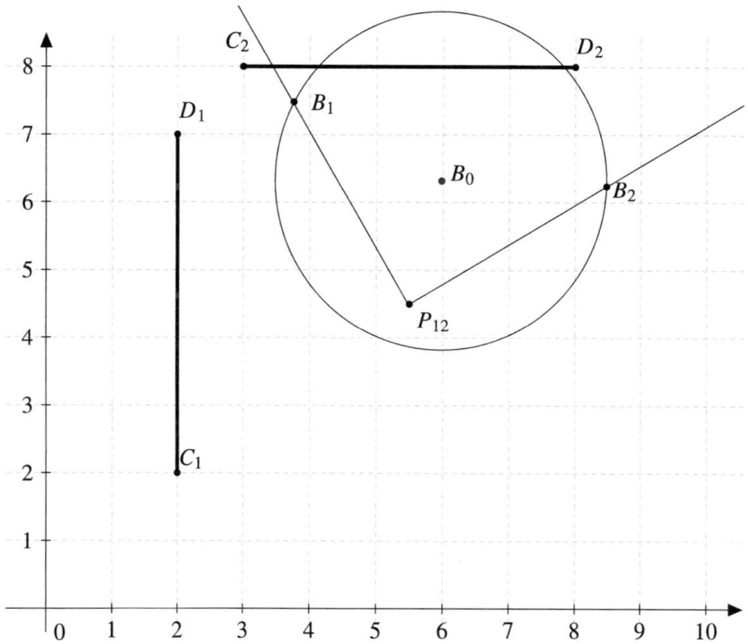

Fig. 7.31 Determining points B_1 and B_2

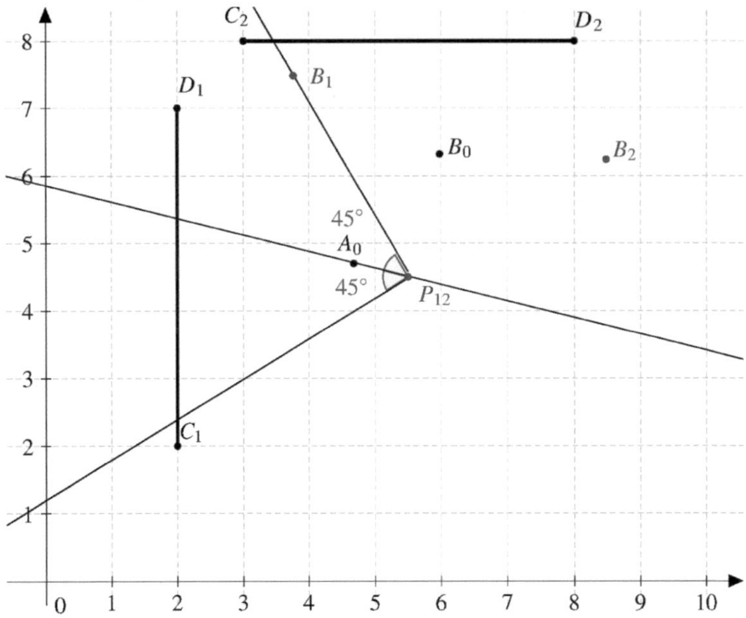

Fig. 7.32 Drawing auxiliary lines

1. Find the pole P_{12}.
2. Determine the angle θ_{12}.
3. Choose a position for B_0.
4. Draw a line through B_0 and P_{12}.
5. Draw the auxiliary rays starting from B_0 and at $\pm\theta_{12}/2$ of the line $\overline{P_{12}B_0}$.
6. Draw a circle centered on B_0 and with the desired radius for the size of the output bar.
7. Determine B_1 and B_2 from the intersection of the circle and the rays.
8. Choose a position for A_0.
9. Draw the line $\overline{P_{12}A_0}$.
10. Draw the auxiliary rays starting from A_0 and at $\pm\theta_{12}/2$ of the line $\overline{P_{12}A_0}$.
11. Draw a circle centered on A_0 and with the desired radius for the size of the input bar.
12. Determine A_1 and A_2 from the intersection of the circle and the rays.
13. Do Filemon's line analysis to check if A_1 is in an allowed region.

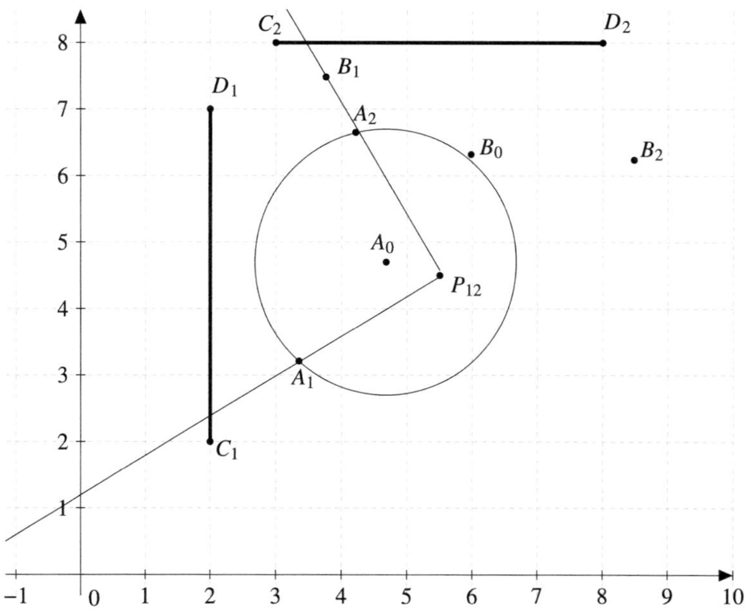

Fig. 7.33 Determining points A_1 and A_2

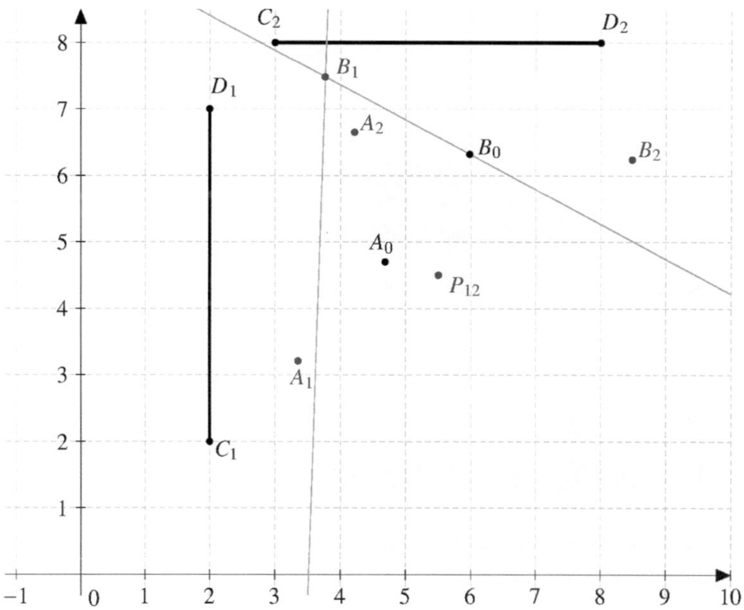

Fig. 7.34 Filemon's lines analysis

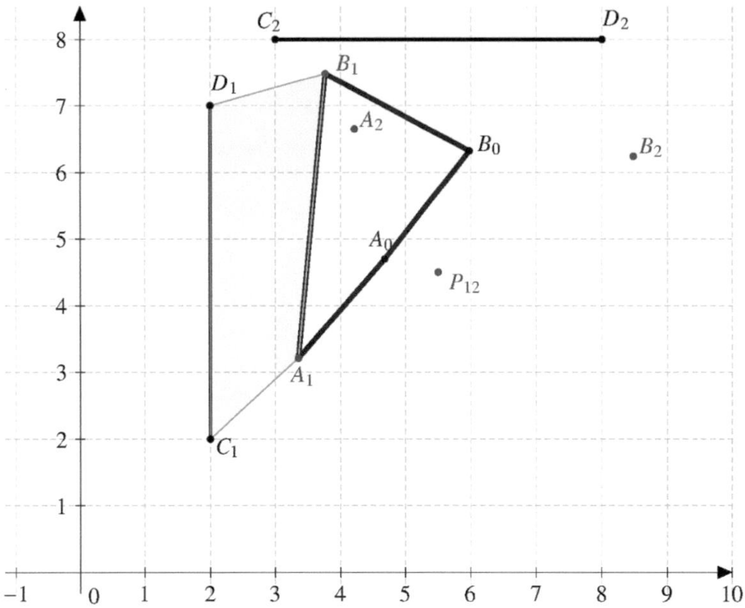

Fig. 7.35 Four-bar mechanism

7.2.5 *Fast Method for 2 FSP Synthesis*

A quick and intuitive method for 2 FSP synthesis is to place A_1 and B_1 directly on C_1 and D_1. And on the medians of C_1C_2 and D_1D_2 choose A_0 and B_0. Although this is a quick and easy method for 2 FSP synthesis, it is not possible to choose the location of the moving pivots. Thus, in the methods presented in Sects. 7.2.3 and 7.2.4 one has three degrees of freedom to adjust the mechanism, whereas in this fast method one only has one degree of freedom. Furthermore, this method does not avoid any of the defects that will be presented below. Therefore, although practical for sketching the mechanism, this method should be used with caution.

7.3 Defects of Dimensional Synthesis

The synthesis of *motion generation*, whether for two, three or four positions, presents some problems. The procedure discussed in this book only ensures that the mechanism can be assembled in all positions, but it does not guarantee operation from one position to the next.

The defects in this synthesis are divided into: circuit defect, branching defect, order defect, low transmission quality defect and workspace constraint defect.

Fig. 7.36 Circuit defect

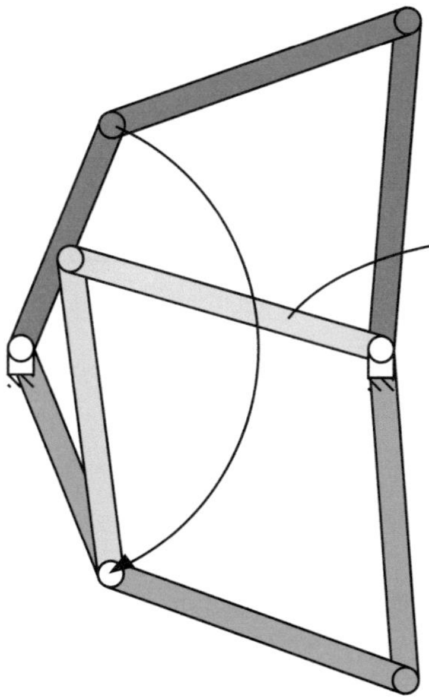

7.3.1 Circuit Defect

The **circuit defect** occurs when the mechanism needs to be disassembled and reassembled in a different *working mode* in order to achieve the desired design positions. This defect was seen in Sect. 7.2.1, and in Sect. 7.2.2 a method was seen to avoid this defect. Another example is shown in the Fig. 7.36, which shows the mechanism mounted in positions 1 and 2. However, starting from position 1, you reach end position 1, not position 2.

In general, Filemon's line analysis avoids circuit defect. However, if the mechanism generated is a Grashof double-rocker, Filemon's line method may fail. However, since the Grashof double-rocker is a specific mechanism that is rarely obtained, the method can be considered robust to avoid the circuit defect. At the end of the synthesis, it is recommended to determine the classification of the mechanism and, if it is a Grashof double-rocker, to carry out a simulation to see if the circuit defect occurs or not.

Mechanisms with circuit defects should be discarded. If the dimensional synthesis has resulted in such a mechanism, the dimensional synthesis should be redone, choosing new positions for the pivots.

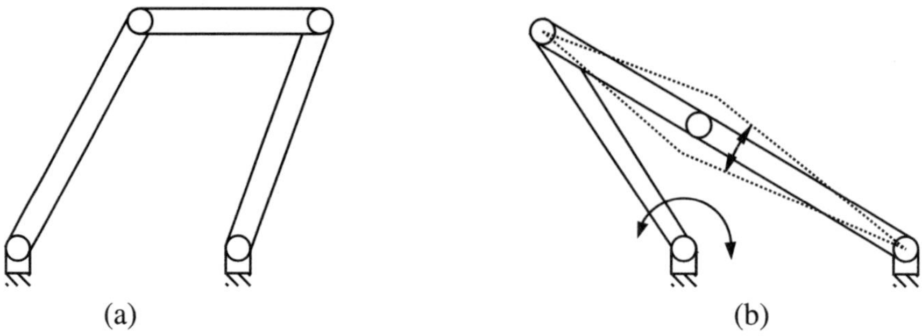

Fig. 7.37 Alignment of coupler and output bars

7.3.2 Branching Defect

The **branching defect** occurs when the mechanism moves from one position to another and during the motion the coupler and output bars align. A mechanism with this defect must be actuated at pivot B_0 to get out of alignment, as mentioned in Sect. 2.4. The branching defect can be avoided by analyzing Filemon's lines, as explained in Sect. 7.2.2. Again, if the resulting mechanism is a Grashof double-rocker, it is necessary to simulate the operation of the mechanism to ensure that alignment does not occur (Fig. 7.37).

7.3.3 Order Defect

Order defect occurs when the moving plane is able to pass through the desired positions, but it does not do so in the desired order. For example, instead of the mechanism taking the moving plane in the following order $1 \rightarrow 2 \rightarrow 3$, it takes it in the following order $1 \rightarrow 3 \rightarrow 2$.

In the 2 FSP synthesis, this defect is not detrimental, as it is enough to reverse the mechanism drive so that instead of the plane going from 2 to 1, it goes from 1 to 2. In the 3 FSP synthesis, it is not always possible to invert the actuation to obtain the correct order. For example, if the mechanism describes the order $1 \rightarrow 3 \rightarrow 2$, reversing it results in $2 \rightarrow 3 \rightarrow 1$. Another alternative would be to start from position 1 and rotate the input link until it reaches position 2 and then 3. However, this would require the input bar to be a crank, which is not always the case. In general, when there is an order defect in the 3 FSP synthesis, the synthesis must be redone by repositioning the pivots.

7.3.4 Low Transmission Quality Defect

Low transmission quality defect occurs when the mechanism has values outside the recommended transmission quality, see Sect. 3.3. Transmission quality is particularly important when the output link has large forces or torques to overcome.

If this defect occurs, the positioning of the pivots must be changed accordingly so that the transmission quality is within the acceptable range.

7.3.5 Workspace Constraint Defect

Workspace constraint defect occurs when external objects influence the operation of the mechanism, leading to collisions.

In this case, several solutions are possible. One can change the shape of the links while maintaining the position of the pivots; one can change the position of the pivots in such a way as to change the dimensions of the bars so that there is no longer a collision; if possible, one can relocate the external object causing the collision.

Sometimes this defect is desired in order to limit the mechanism's operation or to make it function as a lock (*toggle*). An example is folding tables, which have a four-bar mechanism with a limiter, see Fig. 7.38a. When the table is activated, the coupler bar will rotate until it reaches the alignment between the coupler and the output bar (the input bar is the table itself). When it passes this point, due to inertia, the coupler bar tends to continue rotating counterclockwise. When a force is applied to close the table, the coupler will tend to rotate counterclockwise, but due to the limiter, it will not rotate, as shown in Fig. 7.38b. In this case, one wants the mechanism to have a branching defect as well as a workspace constraint.

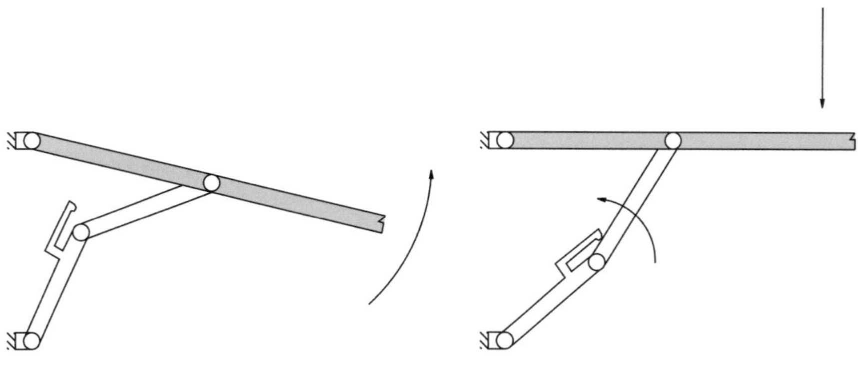

(a) (b)

Fig. 7.38 Four-bar mechanism with limiter. **a** Free motion. **b** limited motion

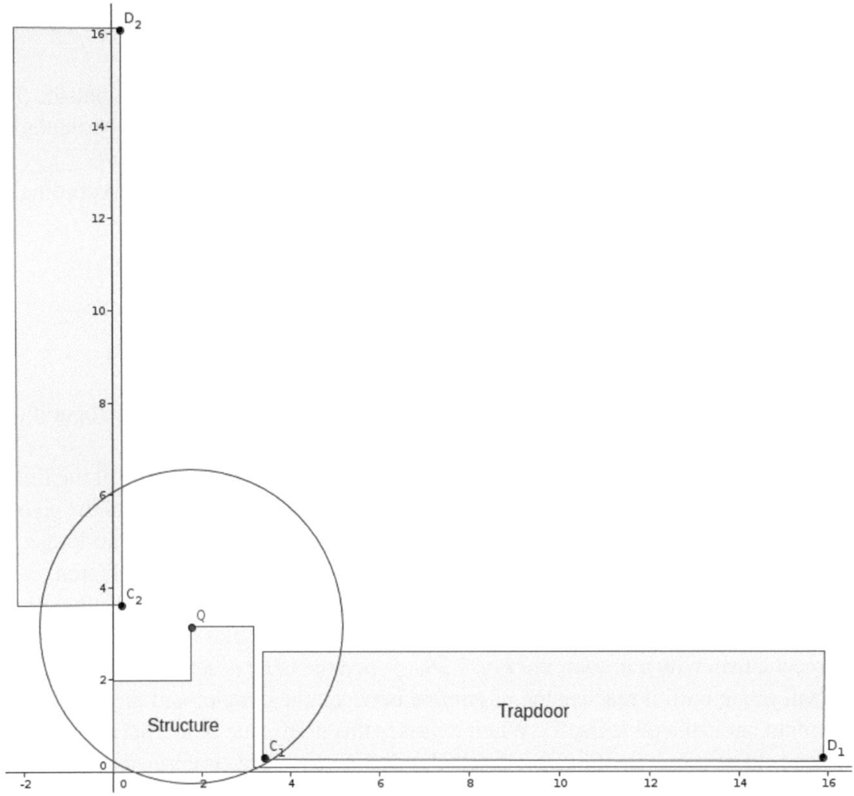

Fig. 7.39 Synthesis for trapdoor mechanism

Problems

7.1 Repeat the example of placing the box lid **using a compass, set squares and ruler**, specifying the moving pivot and then the fixed pivot. Classify the mechanism as Grashof or non-Grashof. Evaluate the transmission quality.

7.2 What happens to the angle ψ_{12} when you change the choice of B_0 along the mediatrix of B_1 and B_2? Evaluate using GeoGebra.

7.3 It is desired to design a four-bar mechanism to open a trapdoor. Figure 7.39 shows position 1 (trapdoor closed) and position 2 (trapdoor open). The hatched area represents part of the structure that the trapdoor cannot enter to avoid collision. The region allowed for the fixed pivots is delimited by the circle, centered on Q and of radius 3.4 $l.u.$ Disregard any collision between the input and output links with the structure or trapdoor (such collision can be resolved by inserting the links in parallel planes other than the structure or trapdoor planes). The trapdoor has a length of 12.5 $l.u.$ and a height of 2.4 $l.u.$ The coordinates of the points in Fig. 7.39 are shown below.

- $C_1 = (3.4; 0.3)$
- $D_1 = (15.9; 0.3)$
- $C_2 = (0.2; 3.6)$
- $D_2 = (0.2; 16.1)$
- $Q = (1.8; 3.1)$.

Using GeoGebra, determine the size of the bars and evaluate the angle of transmission (also using GeoGebra). Note: be careful when choosing the location of the pivots to avoid a circuit defect.

References

1. Norton, R.L.: Design of Machinery: An Introduction to the Synthesis and Analysis of Mechanisms and Machines. McGraw-Hill, New York (2008)
2. McCarthy, J., Soh, G.: Geometric Design of Linkages. Springer, New York (2010)
3. Suh, C., Radcliffe, C.: Kinematics and Mechanisms Design. R.E. Krieger Pub. Co., Malabar (1978)
4. Hartenberg, R.S., Denavit, J.: Kinematic Synthesis of Linkages. McGraw-Hill, New York (1964)
5. Jazar, R.N.: Vehicle Dynamics. Springer, New York (2008)
6. Back, N., Ogliari, A., Dias, A., Silva, J.C.: Projeto integrado de produtos: planejamento, concepção e modelagem. Manole, Barueri (2008)
7. Balli, S.S., Chand, S.: Defects in link mechanisms and solution rectification. Mech. Mach. Theory **37**(9), 851–876 (2002)
8. Ceccarelli, M.: Distinguished Figures in Mechanism and Machine Science: Their Contributions and Legacies. Springer, Netherlands, Dordrecht (2009)

Chapter 8
Dimensional Synthesis of a Four-Bar Linkage for Three Finitely Separated Positions (3 FSP)

Abstract Building on the previous chapter, this one extends dimensional synthesis to accommodate three finitely separated positions (3 FSP) of the moving link. The key concept introduced is the pole triangle, constructed using rotation poles, which provides a geometric foundation for determining fixed and moving pivots. The chapter explores both intuitive graphical methods and systematic geometric constructions, such as using circumcircles to identify link lengths. The synthesis process becomes more constrained as each additional position adds geometric restrictions, limiting the degrees of freedom and increasing the design complexity. Strategies for avoiding mechanism defects and ensuring satisfactory transmission angles are discussed. Worked examples, including the modification of previous 2 FSP problems, help reinforce the methodology. The chapter also offers a "fast method" for quick designs using geometric intersections. This treatment emphasizes both theoretical depth and practical design utility, making it essential for intermediate-level kinematic designers.

Keywords Synthesis for three finitely separated positions · Pole triangle · Circle diagram · Fixed pivot specification · Moving pivot specification · Filemon's line analysis

In this chapter we look at the dimensional synthesis of motion generation for a four-bar mechanism using three finitely separated positions (3 FSP). Concepts that were presented in Chap. 7 are used, so the reader is advised to review concepts from that chapter if necessary.

8.1 Concepts About the 3 FSP Synthesis

In this section we look at concepts about dimensional synthesis for three finitely separated positions.

D. Martins and E. H. Murai, *Mechanisms–Synthesis and Analysis with Applications in Robotics*, Mechanisms and Machine Science 183,
https://doi.org/10.1007/978-3-031-95187-9_8

8.1.1 Poles Triangle

In the 2 FSP synthesis it was seen that there is a point around which it is possible to rotate the moving plane in such a way as to take it from position 1 to position 2. This point was the rotation pole P_{12}. In the three-position synthesis there are three rotation poles, P_{12}, P_{23} and P_{13}.

Pole $\mathbf{P_{12}}$: point around which the moving plane must be rotated by θ_{12} degrees in order to guide the moving plane from position 1 to position 2.
Pole $\mathbf{P_{23}}$: point around which the moving plane must be rotated by θ_{23} degrees in order to guide the moving plane from position 2 to position 3.
Pole $\mathbf{P_{13}}$: point around which the moving plane must be rotated θ_{13} degrees in order to guide the moving plane from position 1 to position 3.

These three poles can be connected by line segments, forming a triangle called **pole triangle**. A property of the pole triangle is that the internal angle of the vertices is half the value of the rotation of the corresponding pole. For example, the vertex P_{13} of the triangle has an angle of $\theta_{13}/2$. This concept is shown in Fig. 8.1. In addition, each side of the triangle is named according to the index common to the two poles through which the side passes, e.g. the side that passes through the poles P_{12} and P_{23} is said to be *side 2*.

In addition to the three poles mentioned above, the 3 FSP synthesis has three more poles, known as **image poles**. These poles are obtained by reflecting the vertices of the triangle in relation to the opposite side. Thus, we have:

Image pole P_{23}^{1}: reflection of pole P_{23} around side 1.
Image pole P_{13}^{2}: reflection of pole P_{13} around side 2.
Image pole P_{12}^{3}: reflection of pole P_{12} around side 3.

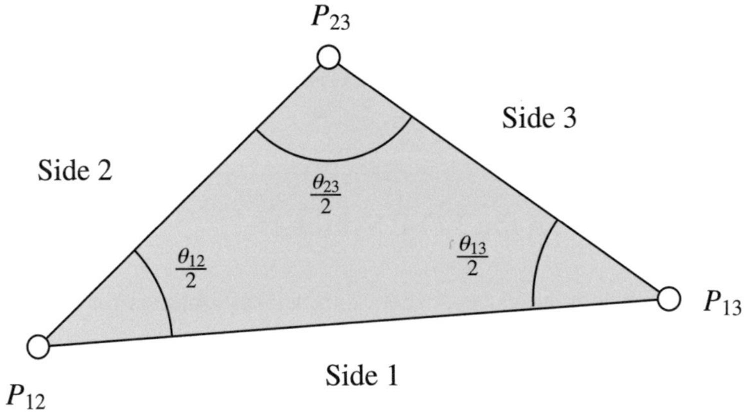

Fig. 8.1 Pole triangle

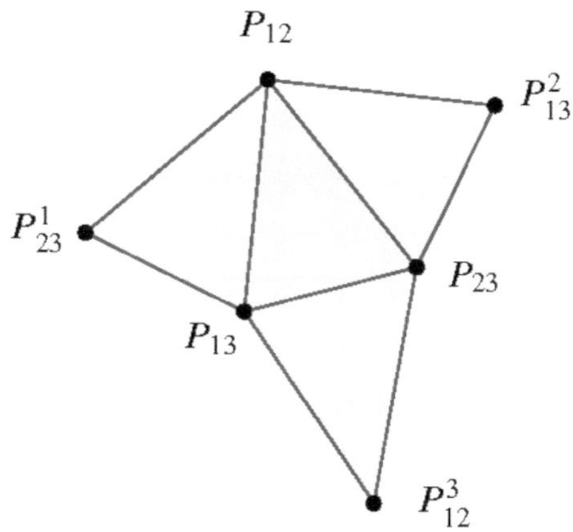

Fig. 8.2 Image poles and image pole triangles

Connecting each image pole with two other original poles results in three more triangles, called **image pole triangles**.

Image pole triangle 1: composed of image pole 1, P_{23}^1, and the other two original poles, P_{12} and P_{13}. Thus, we have $\Delta P_{12} P_{13} P_{23}^1$.
Image pole triangle 2: composed of image pole 2, P_{13}^2, and the other two original poles, P_{12} and P_{23}. Thus, we have $\Delta P_{12} P_{23} P_{13}^2$.
Image pole triangle 3: composed of image pole 3, P_{12}^3, and the other two original poles, P_{13} and P_{23}. Thus, we have $\Delta P_{13} P_{23} P_{12}^3$.

The image poles as well as the image pole triangles are shown in Fig. 8.2.

A motion can look different depending on the reference point used. For example, an observer on Earth would say that the Moon rotates around the Earth, i.e. the center of rotation of the motion is the Earth. On the other hand, the Moon's rotation around its own axis combined with the Moon's translation around the Earth results in always the same side of the Moon facing Earth. Thus, an astronaut on the Moon would say that the Earth stands in the same position in space, it only rotates around its own axis. In this way, different observers position results in different motion perceptions. For the four-bar mechanism, changing the reference point used, i.e. , by choosing another inversion of the kinematic chain, changes the perception of motion. Thus, when the fixed bar is the reference, the pole triangle must be used, but when the coupler bar becomes the reference, the image pole triangles must be used.

Kinematically, the image pole triangle represents the pole triangle when the mechanism is inverted, i.e. the coupler bar becomes the fixed bar and the fixed bar becomes the coupler (the chosen inversion of the kinematic chain is changed). Thus, image pole triangle 1 represents the inversion of the mechanism when it is in position 1. Similarly, image pole triangle 2 represents the inversion of the mechanism when it is in position 2, and the same is true for image pole triangle 3 and position 3.

When the mechanism is inverted in position 1, the new pole triangle obtained is image pole triangle 1. For example, Fig. 8.3a shows a four-bar mechanism in

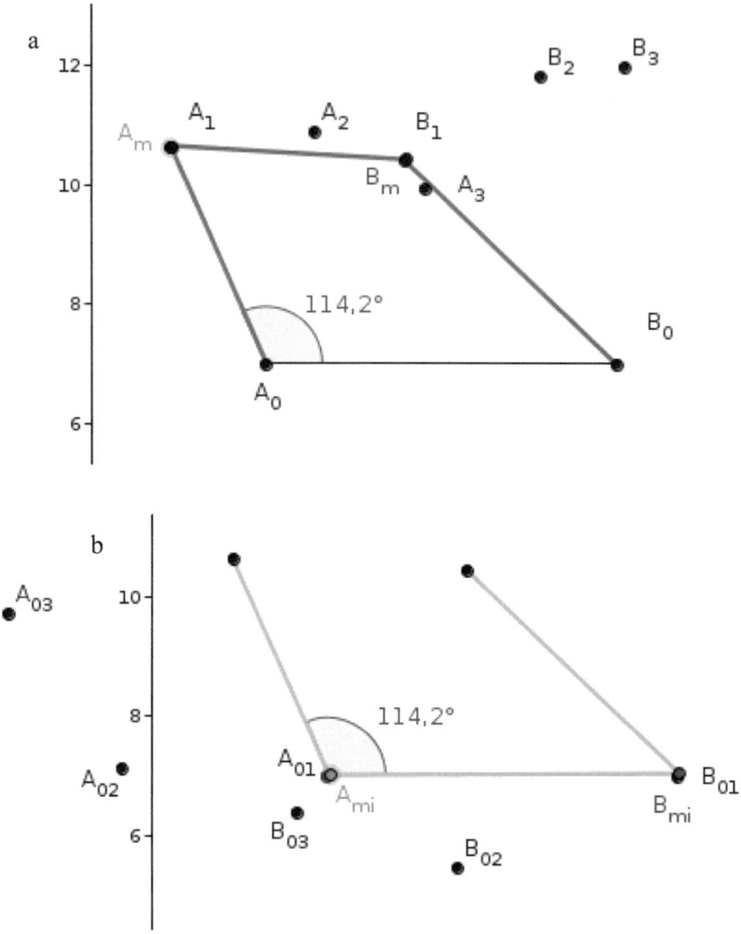

Fig. 8.3 Mechanisms in position 1. **a** Original mechanism. **b** Inverted mechanism

position 1, and Fig. 8.3b shows the other inversion also in position 1. In this case, as the inversion was adopted in position 1, the mechanisms overlap.

When the original mechanism is moved to position 2, the angle between the fixed bar and the input bar becomes 78.21°, as can be seen in the Fig. 8.4a. When the inverted mechanism is actuated so that the angle between the input bar and the coupler is the same 78.21°, the mechanism in Fig. 8.4b is obtained.

Moving the original mechanism to position 3, the angle between the fixed and input bars becomes 47.17°, as shown in Fig. 8.5a. By imposing this same angle between the input and coupler bars of the inverted mechanism, the mechanism in Fig. 8.5b is obtained.

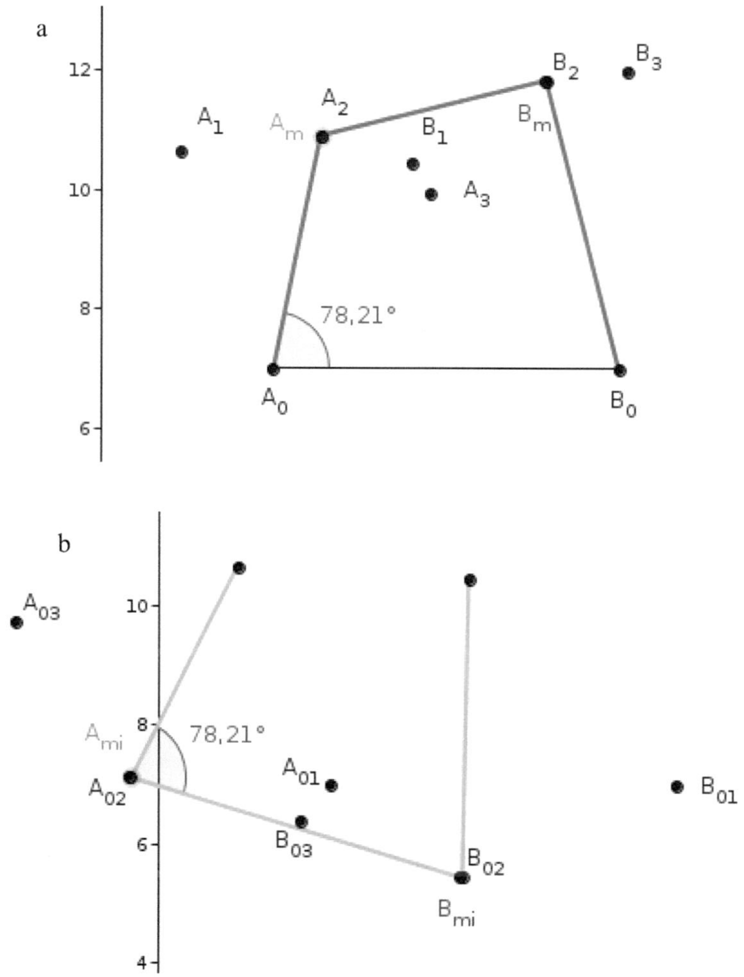

Fig. 8.4 Mechanisms in position 2. **a** Original mechanism. **b** Inverted mechanism

Using poles P_{12}, P_{23} and P_{13} from the original mechanism, one can make the pole triangle, as shown in Fig. 8.6a. Repeating the procedure for the inverted mechanism, the result is the triangle shown in Fig. 8.6b.

By superimposing the two mechanisms, one can see that the pole triangle of the inverted mechanism in position 1 is the image pole triangle 1 of the original mechanism, as shown in Fig. 8.7.

More content about the image pole triangle can be found at [1, 2].

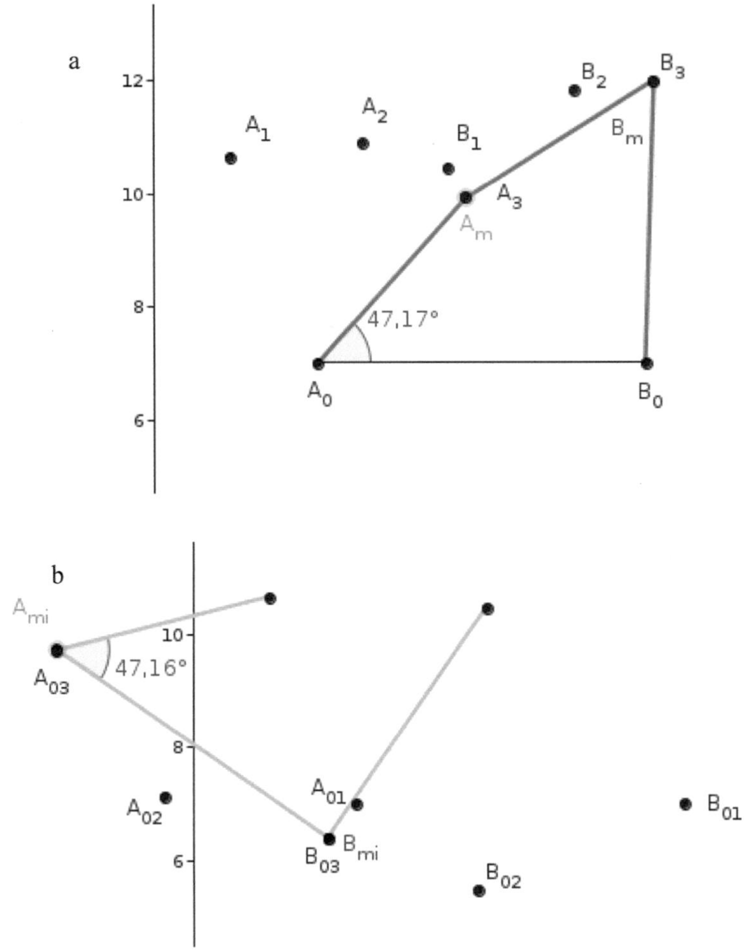

Fig. 8.5 Mechanisms in position 3. **a** Original mechanism. **b** Inverted mechanism

8.1.2 *Filemon's Lines for 3 FSP*

This section aims to complement the concepts presented in Sect. 7.2.2, and the reader should be familiar with Filemon's lines analysis for two finitely separated positions.

The principle of the method remains the same as presented in Sect. 7.2.2: find the angle at which the coupler bar rotates around the output bar during actuation. The forbidden zone is determined so that during this rotation the coupler and output bar do not align, i.e. the pivot A_m does not cross the **line** $\overline{B_0B_m}$. However, for 3 FSP synthesis, the coupler bar can reverse the direction of rotation when traveling through positions 1, 2 and 3. For example, when going from position 1 to position 2, the coupler bar rotates clockwise in relation to the output bar (ψ_{12} is negative) and

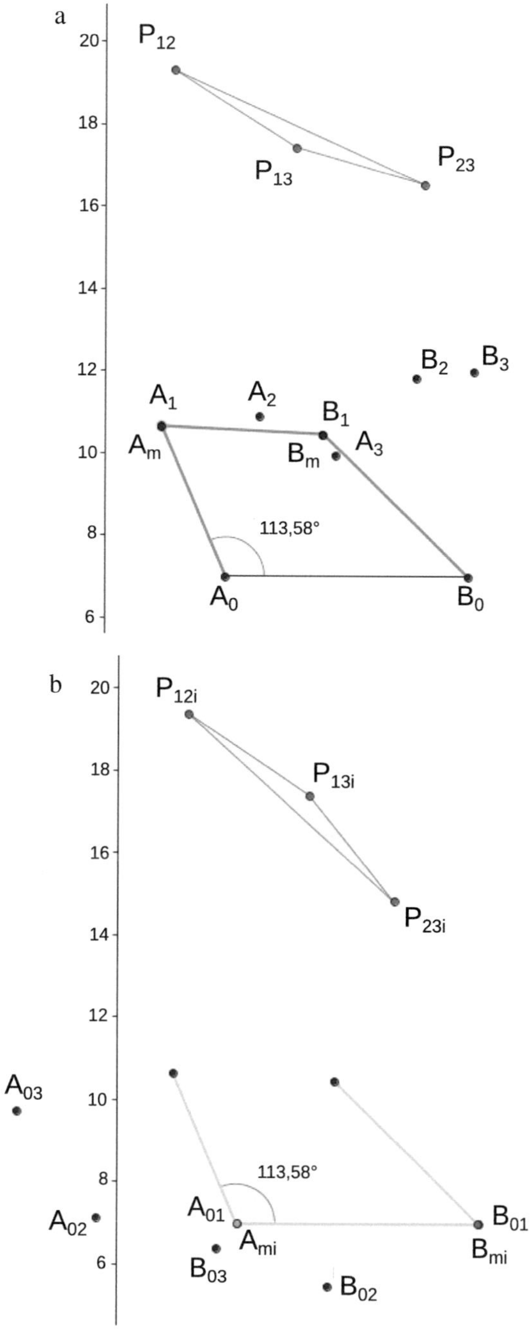

Fig. 8.6 Pole and image pole triangles. **a** Original mechanism. **b** Inverted mechanism

Fig. 8.7 Original and inverted mechanisms superimposed

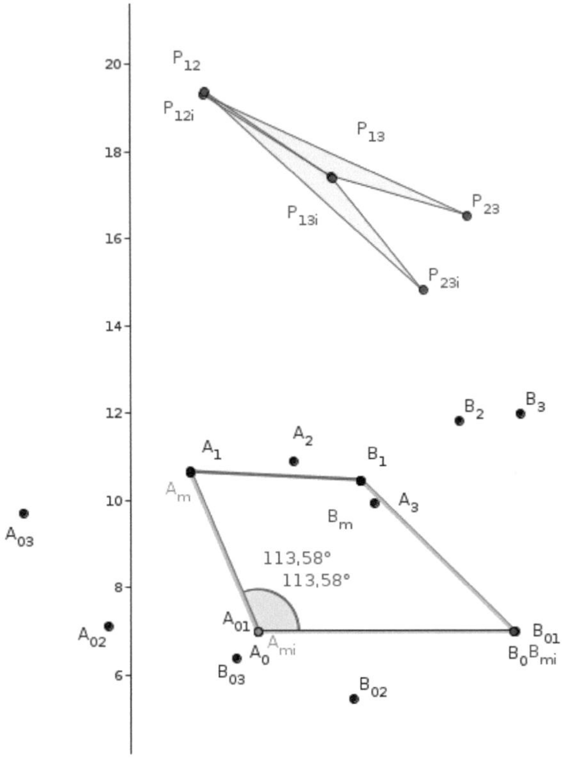

then, to go from position 2 to position 3, the coupler bar rotates counterclockwise (ψ_{23} is positive). Thus, more attention is needed when doing the Filemon's line analysis for 3 FSP.

Determining the angles ψ_1, ψ_2 and ψ_3 remains the same as for 2 FSP, just repeating the procedure once again for the third position. The angles $\psi_{12} = \psi_2 - \psi_1$ and $\psi_{13} = \psi_3 - \psi_1$ are calculated. Depending on the values of ψ_{12} and ψ_{13}, there are four possibilities, as shown in Fig. 8.8.

Figure 8.8 shows that ψ_{12} and ψ_{13} can be positive or negative, affecting the forbidden region. Also, in this figure, the point B_m is used to generically represent the points B_1, B_2 and B_3 in positions 1, 2 and 3.

8.1.3 Circle Diagram

The circle diagram is a procedure for determining the forbidden region for positioning B_1 in order to avoid the branching problem. Using the image pole triangle 1, three circles are made, each centered on one of the sides of the triangle and with a diameter equal to its side. The forbidden regions are those in which there are an odd number

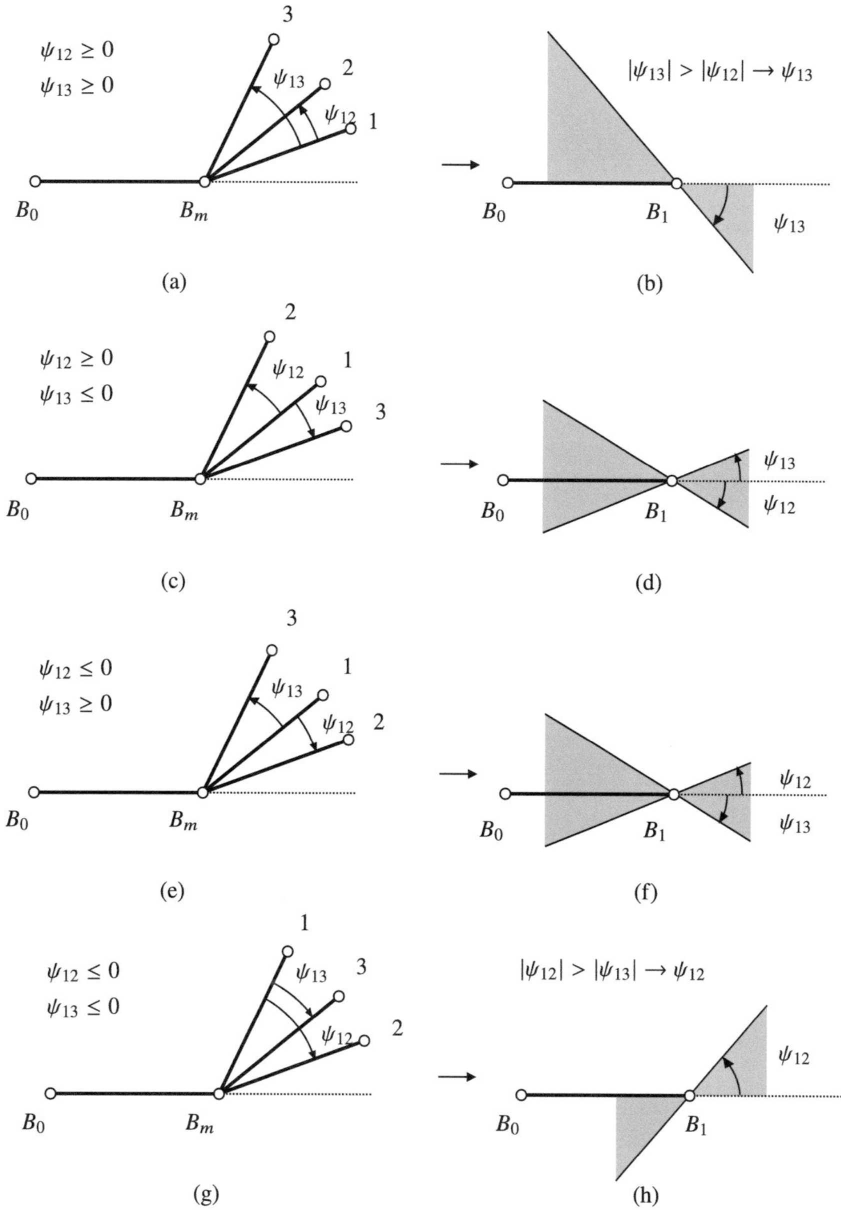

Fig. 8.8 Filemon's line analysis for 3 FSP

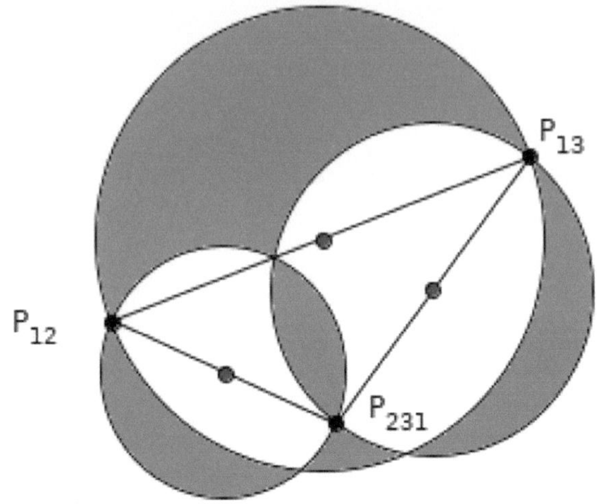

Fig. 8.9 Diagram of circles with the forbidden region for B_1 highlighted

of circles. An example is shown in Fig. 8.9. If one wants to determine the forbidden region for B_2, the procedure would be the same, but using the image pole triangle 2 and, analogously, the same for B_3. However, it is only needed to draw the circle diagram for one of the image poles.

8.2 Synthesis of Three Finitely Separated Positions

The dimensional synthesis of the four-bar mechanism with 3 finitely separated positions, or *3 FSP synthesis*, aims to determine the dimensions of a four-bar mechanism so that the coupling link does a motion that passes through three desired positions.

8.2.1 Specifying Moving Pivots

The three desired positions are shown in Fig. 8.10, the problem is similar to the one presented for two positions, but one intermediate position has been inserted.

Initially, the three rotation poles must be determined using the mediatrices, as explained in the 2 FSP synthesis. Thus, the pole of rotation P_{12} is given by the crossing of mediatrices C_1C_2 and D_1D_2. This pole is shown in Fig. 8.11. The pole of rotation P_{23} is given by the crossing of mediatrices of C_2C_3 and D_2D_3, as shown in Fig. 8.12. The pole of rotation P_{13} is given by the crossing of mediatrices C_1C_3 and D_1D_3, as shown in Fig. 8.13.

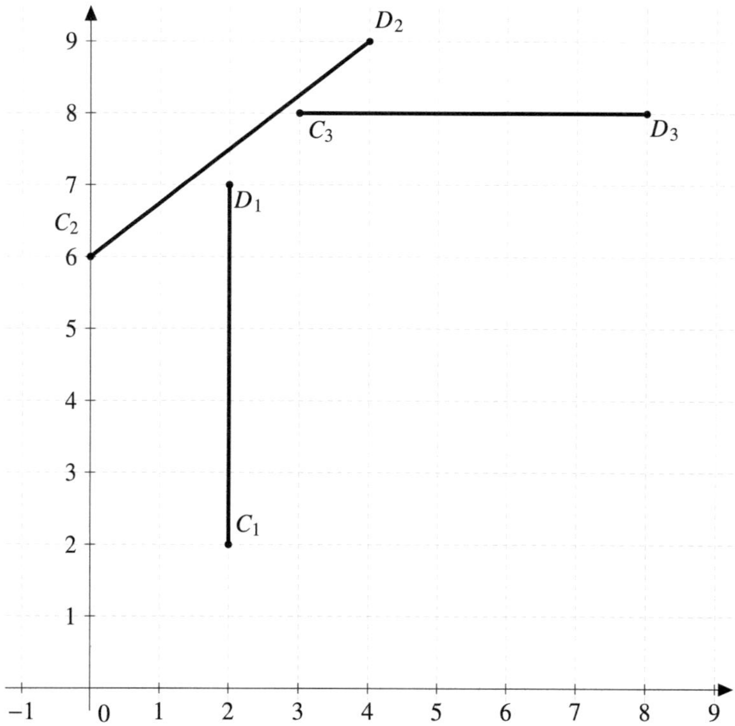

Fig. 8.10 Desired positions

Figure 8.14 shows the pole triangle ($\Delta P_{12}P_{23}P_{13}$) and the image pole triangles ($\Delta P_{12}P_{13}P_{23}^{1}$, $\Delta P_{12}P_{23}P_{13}^{2}$ and $\Delta P_{13}P_{23}P_{12}^{3}$). Using the image pole triangle 1, determine the forbidden region for positioning B_1, as shown in Fig. 8.15. B_1 must be chosen outside this region.

Once B_1 has been chosen, B_2 and B_3 must be determined. There are two ways to determine these points. The first has already been explained in Chap. 7, i.e. rotate B_1 around P_{12} at an angle of θ_{12} and repeat the procedure from position 2 to 3, or even from position 1 to 3.

The second method involves determining the **cardinal point** B_g and using that point to determine B_2 and B_3. To obtain B_g, reflect point B_1 around side 1 of the triangle of poles. Then reflect B_g around sides 2 and 3 of the pole triangle, obtaining points B_2 and B_3 respectively. The cardinal point as well as points B_2 and B_3 are shown in Fig. 8.16. When the synthesis procedure is carried out manually, it is more practical to use the first method. When the synthesis procedure is carried out using GeoGebra, it is more practical to use the second method.

Once the points B_1, B_2 and B_3 are known, it is possible to determine B_0. This fixed pivot is at the intersection of mediatrices B_1B_2 and B_2B_3. Another way to determine

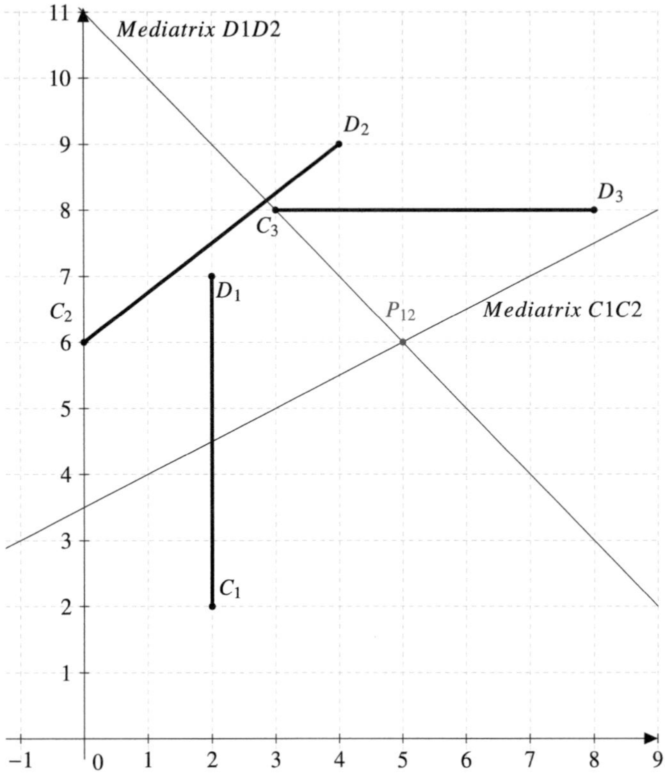

Fig. 8.11 Pole P_{12}

B_0 is to draw a circle that passes through B_1, B_2 and B_3, and the center of this circle is B_0. Figure 8.17 shows how B_0 is obtained.

Filemon's lines analysis is shown in Fig. 8.18. Note that $\psi_1 = 93.02°$, $\psi_2 = 88.21°$ and $\psi_3 = 136.57°$, so we have $\psi_{12} = \psi_2 - \psi_1 = -4.81°$ and $\psi_{13} = \psi_3 - \psi_1 = 43.55°$. Figure 8.19 shows the forbidden region for A_1 as well as the chosen A_1.

Once A_1 has been chosen, A_2 and A_3 must be determined, again using the cardinal point A_g. The points A_g, A_2 and A_3 are shown in Fig. 8.20. At the intersection of mediatrices A_1A_2 and A_2A_3 is the fixed pivot A_0. These points are represented in Fig. 8.21. The final mechanism is shown in Fig. 8.22.

Two degrees of freedom are used to determine the output bar. These two degrees lie in the choice of the moving pivot B_1, because when this pivot is positioned, the coordinates x and y are chosen (choice of a point on the plane). The same applies to the input bar. Thus, compared to the 2 FSP synthesis, inserting an intermediary position adds a constraint to the design and consequently reduces one degree of freedom.

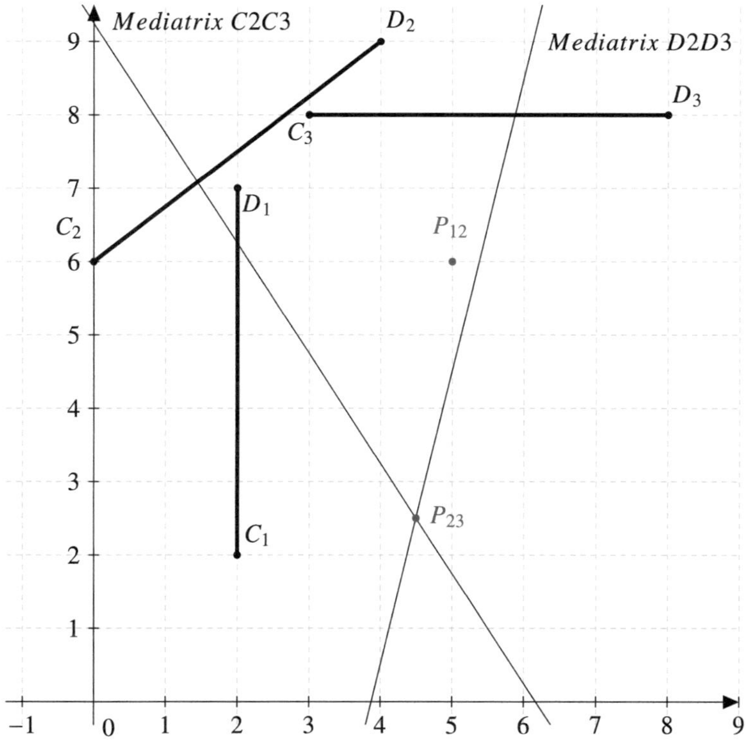

Fig. 8.12 Pole P_{23}

The procedure is summarized below:

1. Determine the poles and the image pole 1.
2. Determine the poles triangle and the **image pole triangle 1**.
3. Determine the forbidden region for B_1 by analyzing the circle diagram using the image pole triangle 1.
4. Choose B_1 outside the forbidden region.
5. Determine B_g and then B_2 and B_3 using the **pole triangle**.
6. Draw the mediatrices of B_1B_2 and B_2B_3, and at their intersection insert B_0.
7. Determine the forbidden region for A_1 by analyzing Filemon's lines.
8. Choose A_1 outside the forbidden region.
9. Determine A_g and then A_2 and A_3 using the **poles triangle**.
10. Draw the mediatrices of A_1A_2 and A_2A_3, and insert A_0 at their intersection.

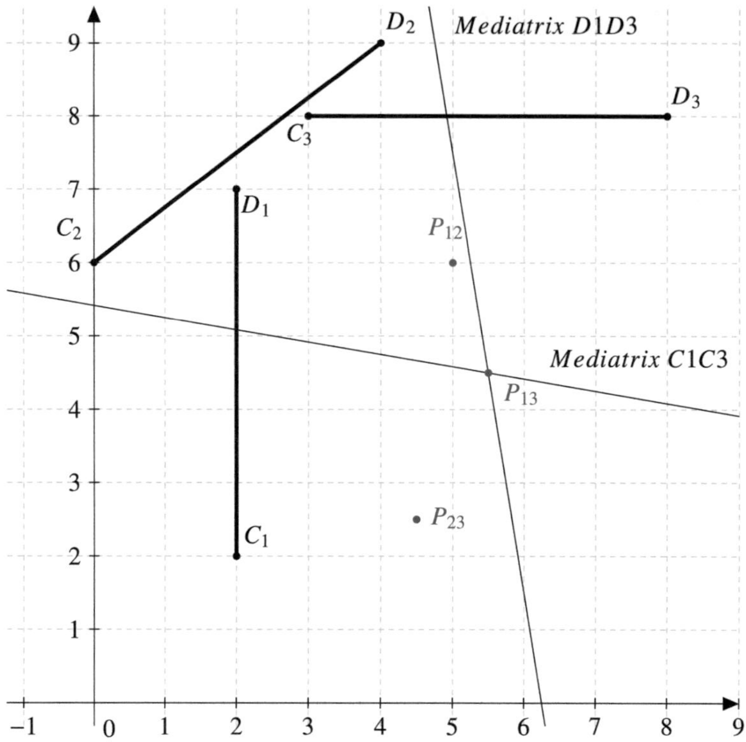

Fig. 8.13 Pole P_{13}

8.2.2 *Specifying Fixed Pivots*

To specify the fixed pivots in the 3 FSP synthesis, it is used another inversion of the four-bar mechanism. When the coupler bar is fixed, the bar that was previously fixed becomes the coupler, the fixed pivots become moving and the moving pivots become fixed. So when specifying the moving pivots of the inverted mechanism, one is actually specifying the fixed pivots of the initial mechanism.

It is possible to work with the inverted mechanism in position 1, 2 or 3. In this text we will only work with the mechanism for the first position, but the procedure for the others positions is analogous. The previously fixed pivot B_0 will be B_{01}, B_{02}, B_{03} or B_{0m}, depending on whether the mechanism is in position 1, 2, 3 or in between, respectively. The same nomenclature follows for the previously fixed pivot A_0.

Starting from the three positions, the image pole triangle 1 is determined, as shown in Fig. 8.23. Next, choose the location of the fixed pivot, B_{01}. The cardinal point B_{0g} must be determined by reflecting B_{01} around side 1 of the **image pole triangle 1**. To determine B_{02} and B_{03}, simply reflect B_{0g} around sides 2 and 3, respectively, of the **image pole triangle 1**. These points are shown in Fig. 8.24.

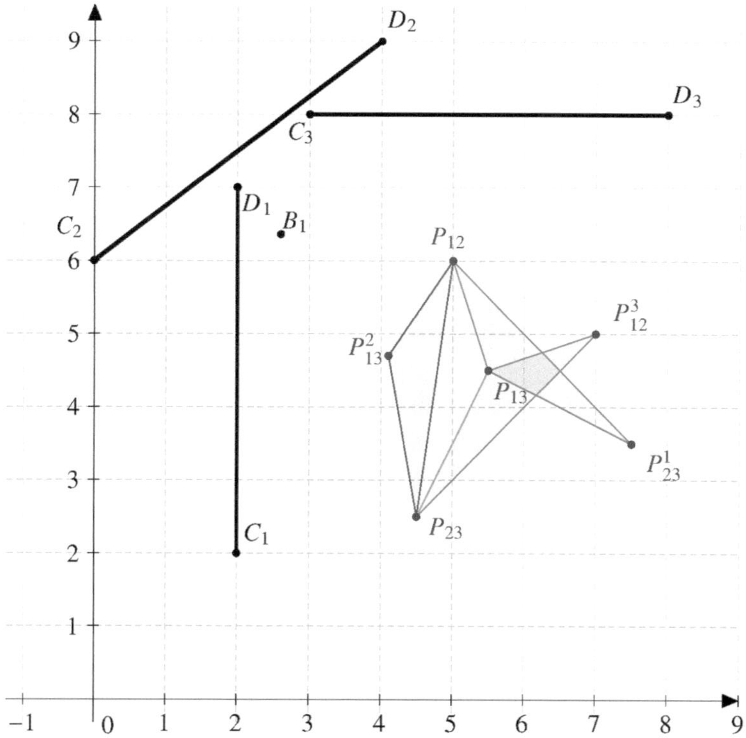

Fig. 8.14 Pole and image pole triangles

The pivot B_1 will be at the meeting of the medians of $B_{01}B_{02}$ and $B_{02}B_{03}$. Another equivalent method for determining B_1 is to draw a circle passing through B_{01}, B_{02} and B_{03}, with B_1 at the center of this circle. The Fig. 8.25 shows how B_1 was obtained. As the inversion used was in position 1 (image pole triangle 1), then the pivot B_0 is given by B_{01}. If position 2 were used, the image pole triangle 2 would have to be used and the point B_0 would be given by B_{02}.

After determining B_1, check whether it is outside the region prohibited by the circle diagram. If B_1 is in a forbidden region, choose another location for B_{01}.

Once B_1 is known, B_2 and B_3 are determined, which can be done by rotating B_1 around pole P_{12} and rotating B_2 around pole P_{23}, respectively. Another option would be to reflect B_1 on the sides of the **pole triangle**, as was seen for the specification of the moving pivots (Sect. 8.2.1). Once points B_0, B_1, B_2 and B_3 are known, it is possible to determine the forbidden region for A_1 by analyzing Filemon's lines.

The procedure carried out for the output link is repeated for the input link, as shown in Figs. 8.26, 8.27 and 8.28. Check that the point A_1 obtained is in the region allowed by Filemon's lines. The mechanism obtained is shown in Fig. 8.29.

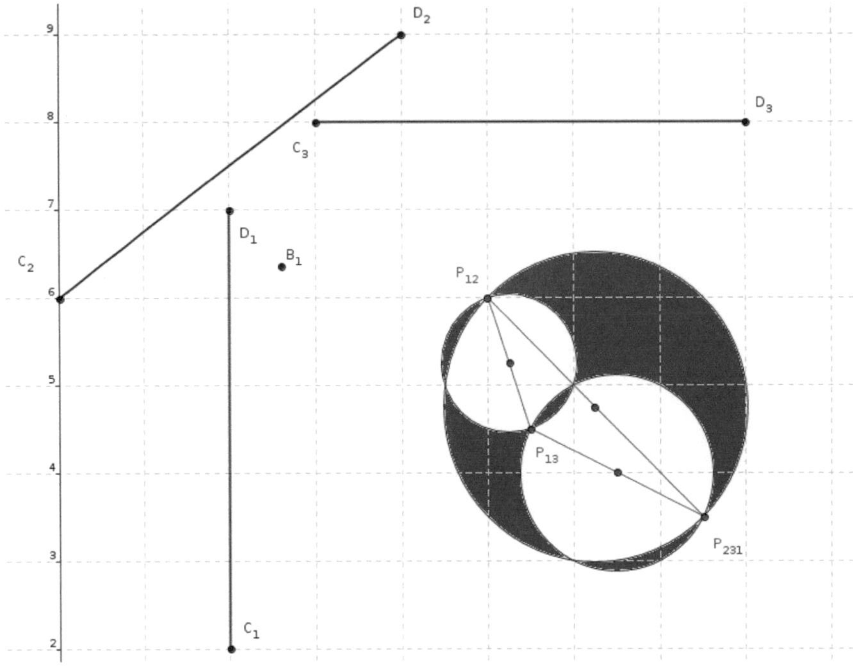

Fig. 8.15 Diagram of circles and choice of B_1

Below is a summary of the procedure:

1. Determine the poles and the image pole 1.
2. Determine the poles triangle and the **image pole triangle 1**.
3. Determine the forbidden region for B_1 by analyzing the circle diagram using the image pole triangle 1.
4. Choose B_{01}.
5. Determine B_{0g} and then B_{02} and B_{03} using the **image pole triangle 1**.
6. Draw the mediatrices of $B_{01}B_{02}$ and $B_{02}B_{03}$, and insert B_1 at their intersection.
7. Check that B_1 is outside the region forbidden by the circle diagram.
8. Determine B_2 and B_3 using the **pole triangle**.
9. Determine the forbidden region for A_1 by analyzing Filemon's lines.
10. Choose A_{01}.
11. Determine A_{0g} and then A_{02} and A_{03} using the **image pole triangle 1**.
12. Draw the mediatrices of $A_{01}A_{02}$ and $A_{02}A_{03}$, and insert A_1 at their intersection.
13. Check that A_1 is outside the region forbidden by Filemon's lines.

More information on the 3 FSP synthesis can be found at [1, 3, 4].

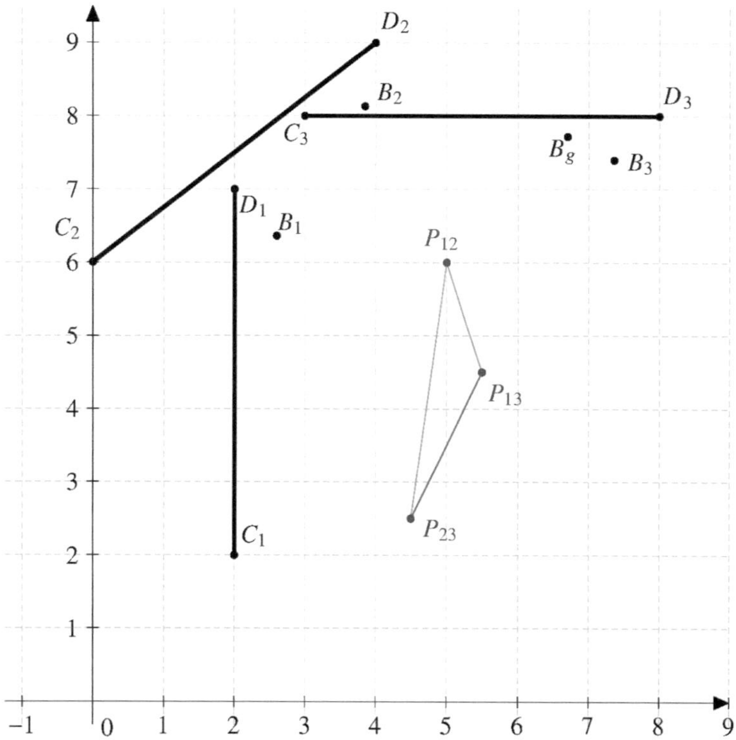

Fig. 8.16 Determination of B_2 and B_3

8.2.3 Fast Method for 3 FSP Synthesis

A fast method for 3 FSP synthesis is to use C_1, C_2 and C_3 as the positions of A_m. So, when a circle that passes through C_1, C_2 and C_3 is drawn, the center of the circle is where A_0 will be. Repeating the procedure with D_1, D_2 and D_3 gives us B_0. The radii of the circles that pass through C_1, C_2 and C_3 and through D_1, D_2 and D_3 represent the dimensions of the input and output bars.

This intuitive method is quick and easy to use both in GeoGebra and manually, but given C_1 and D_1, there is no choice being made, as the positions of A_1 and B_1 have been restricted to those of $A_1 \equiv C_1$ and $B_1 \equiv D_1$. Furthermore, this method does not avoid the defects presented in Sect. 7.3.

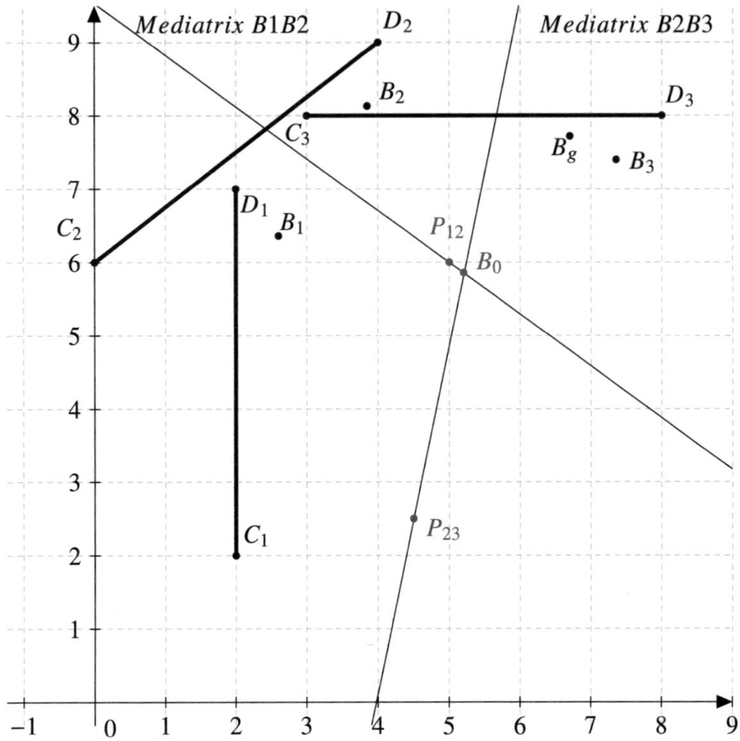

Fig. 8.17 Determination of B_0

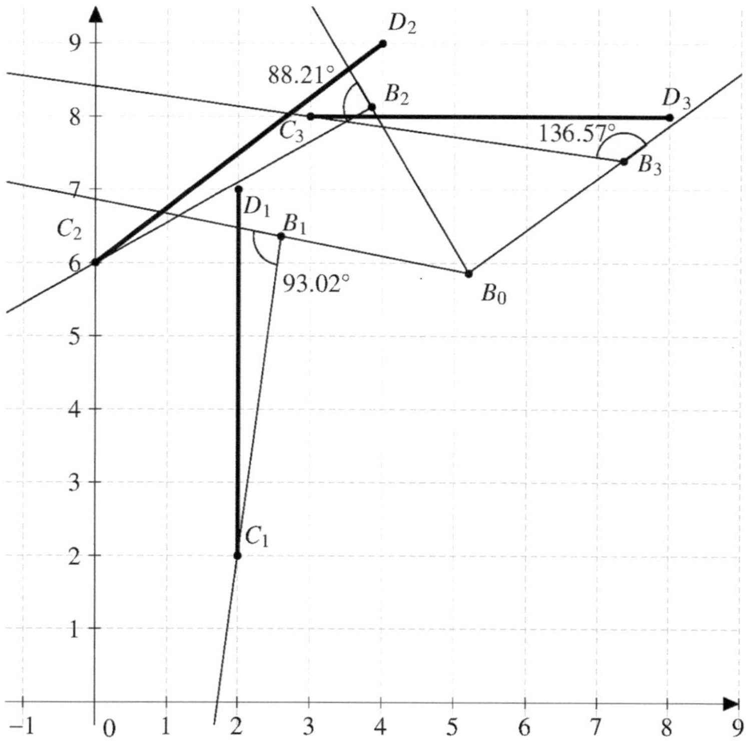

Fig. 8.18 Angles ψ_1, ψ_2 and ψ_3

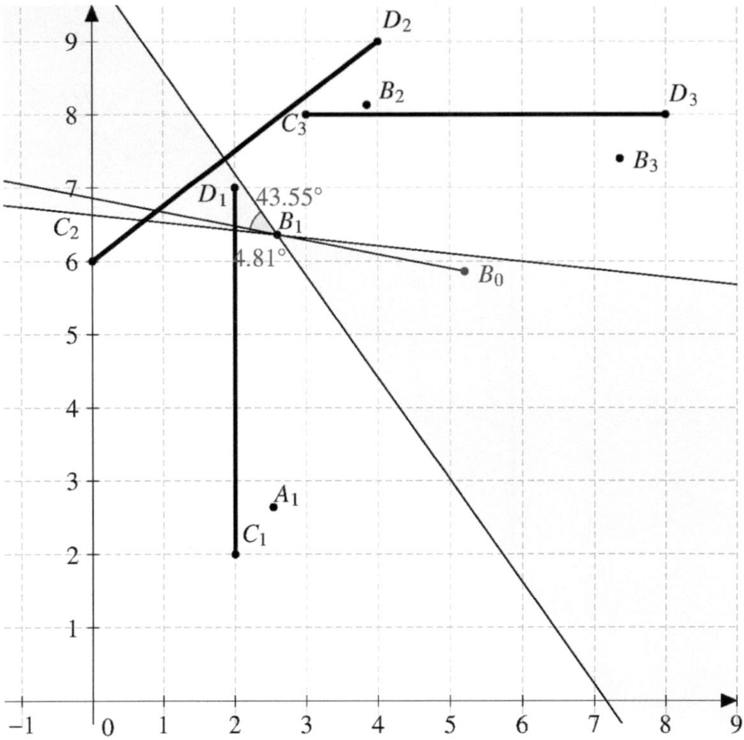

Fig. 8.19 Choice of A_1

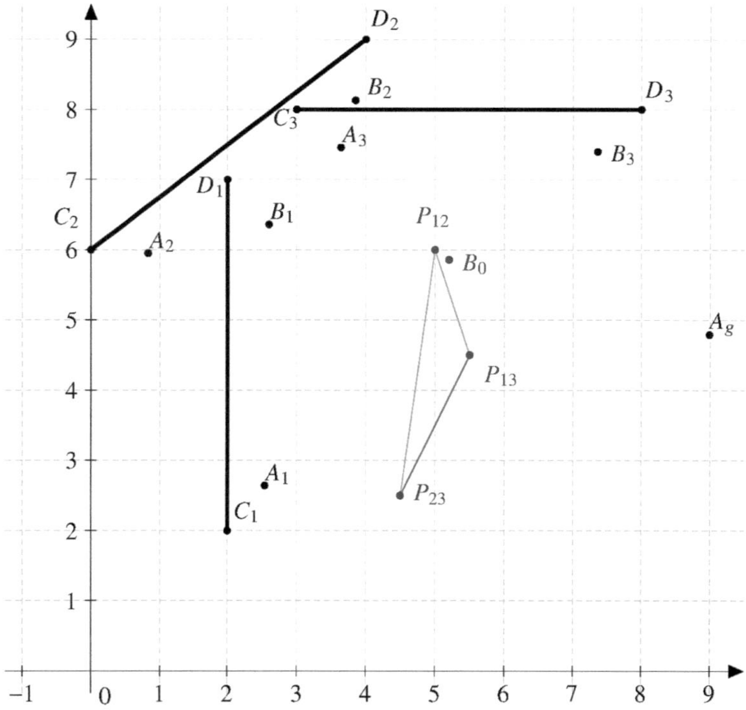

Fig. 8.20 Determining A_2 and A_3

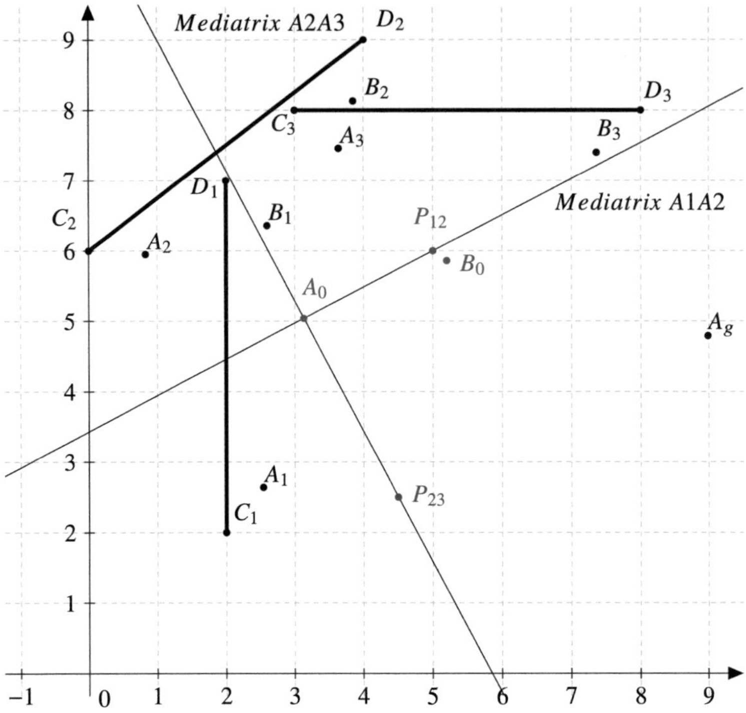

Fig. 8.21 Determining A_0

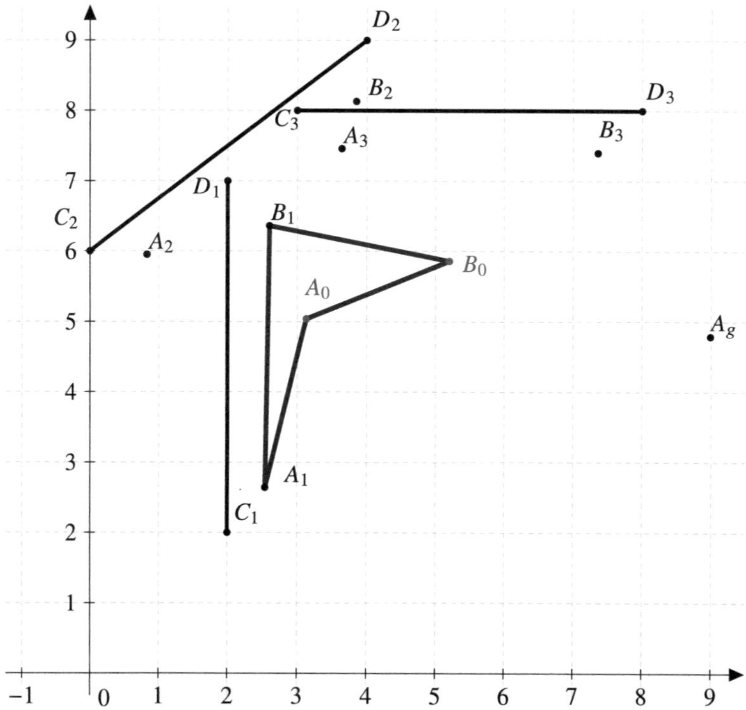

Fig. 8.22 Four-bar mechanism

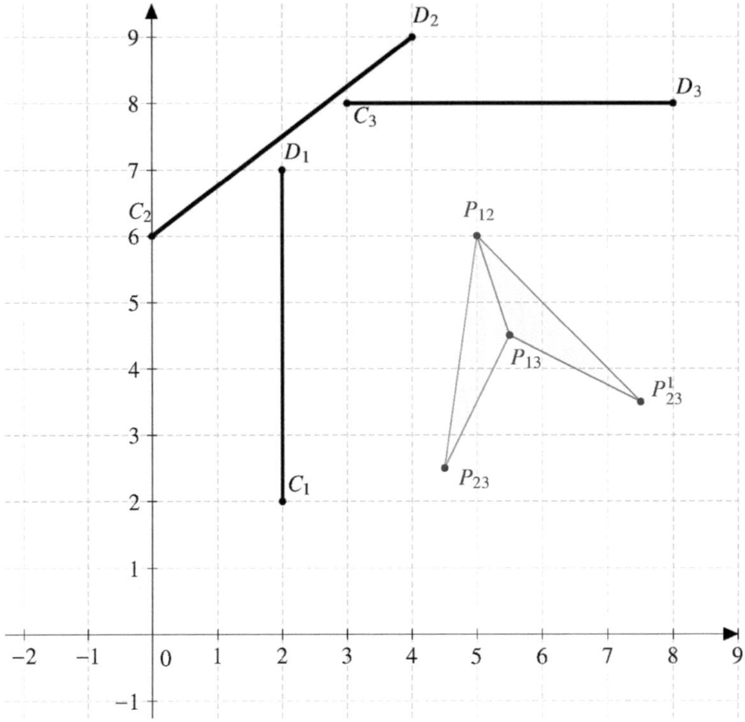

Fig. 8.23 Image pole triangle 1

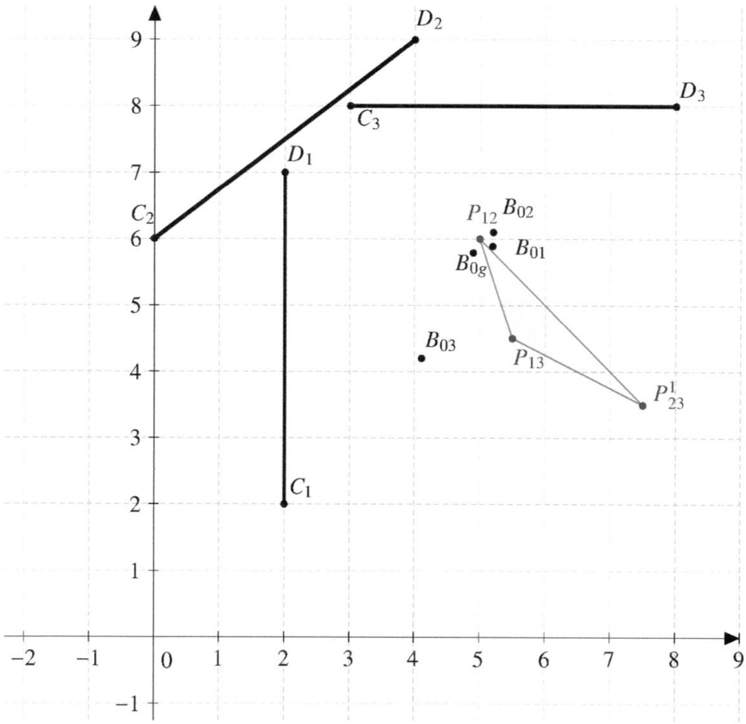

Fig. 8.24 Determining B_{02} and B_{03}

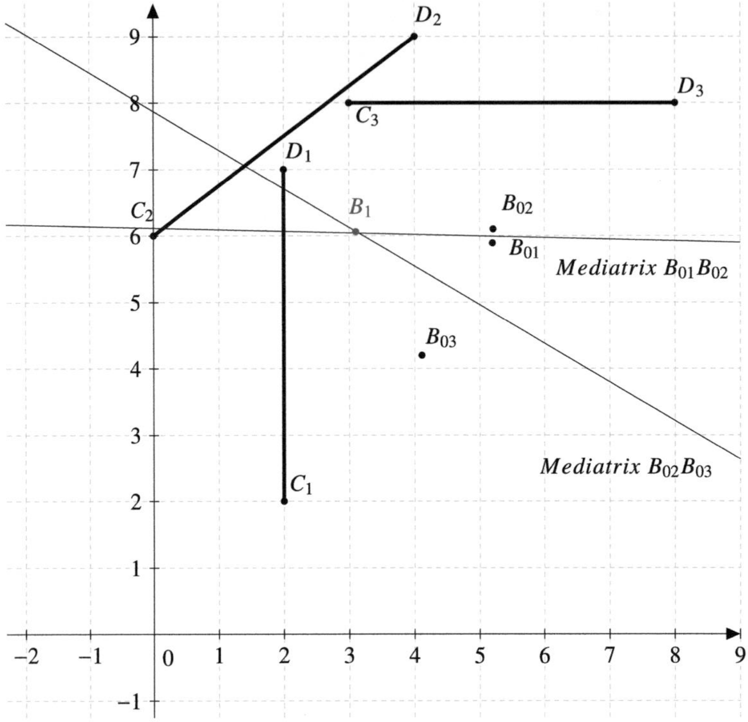

Fig. 8.25 Determining B_1

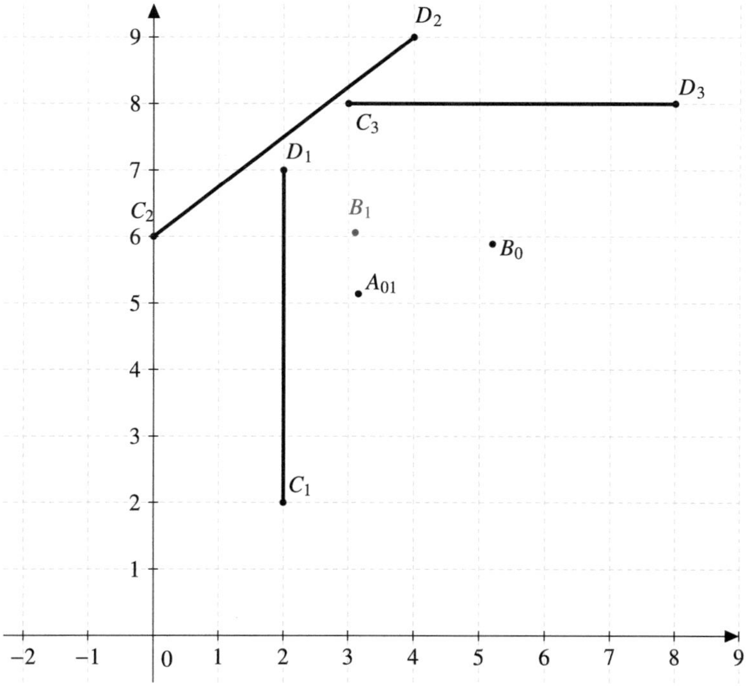

Fig. 8.26 Choice of A_{01}

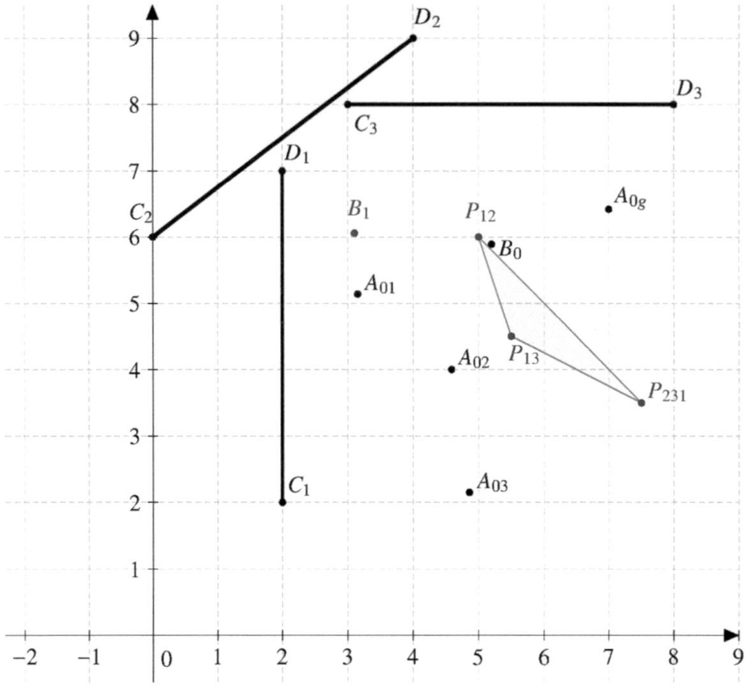

Fig. 8.27 Determining A_{02} and A_{03}

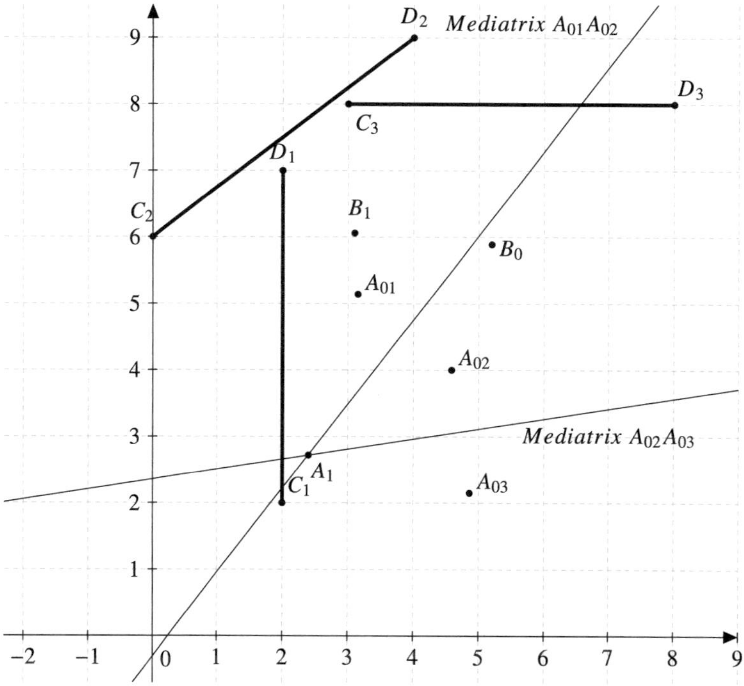

Fig. 8.28 Determining A_1

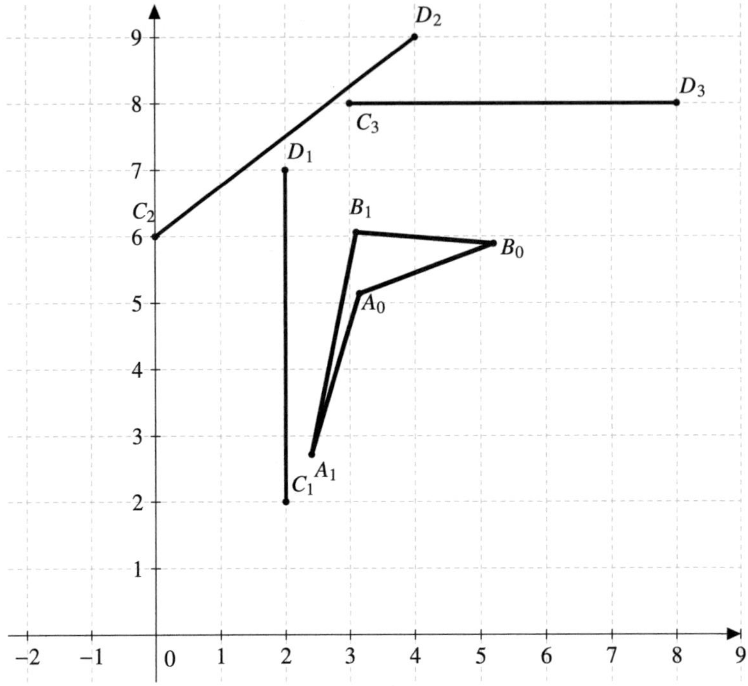

Fig. 8.29 Four-bar mechanism

Problems

8.1 Redo the problem of placing the box lid adding an intermediary position. Use a compass, set squares and ruler. Specify the moving pivot and then the fixed pivot. Classify the mechanism as Grashof or non-Grashof. Evaluate the transmission quality.

8.2 Redo the trapdoor exercise in Chap. 7 by inserting an intermediary position.

References

1. McCarthy, J., Soh, G.: Geometric Design of Linkages. Springer, New York (2010)
2. Hartenberg, R.S., Denavit, J.: Kinematic Synthesis of Linkages. McGraw-Hill, New York (1964)
3. Norton, R.L.: Design of Machinery: An Introduction to the Synthesis and Analysis of Mechanisms and Machines. McGraw-Hill, New York (2008)
4. Hunt, K.H.: Kinematic Geometry of Mechanisms. Clarendon Press, Oxford (1978)

Chapter 9
Function Generation and Crank-rocker Dimensional Syntheses

Abstract This chapter focuses on the dimensional synthesis of four-bar mechanisms specifically for function generation and crank-rocker applications. It revisits Freudenstein's equations, a mathematical framework used to relate input and output angles for precise motion control. By applying trigonometric identities and algebraic manipulations, readers learn how to derive links dimensions that fulfill a specific angular function. The chapter also delves into crank-rocker mechanisms, explaining how to design systems that convert continuous input rotation into controlled oscillatory motion. Practical criteria such as the Grashof condition, transmission quality, and mechanical advantage are discussed to evaluate mechanism viability. Analytical procedures are supplemented by graphical interpretations to enhance intuition. The content bridges theory and application, equipping readers with tools to synthesize mechanisms that accurately perform specified mathematical functions.

Keywords Function generation synthesis · Crank-rocker synthesis

In this chapter we look at the dimensional synthesis of function generation and the dimensional synthesis of the crank-rocker.

9.1 Function Generation

The function generation dimensional synthesis has already been explained in Sect. 7.1.1. In this synthesis, the aim is to determine the dimensions of the four links in order to satisfy a correlation between the angles of the input and output links. The procedure for generating the function is based on Freudenstein's equations[1] [1].

Consider the parameters shown in the four-bar mechanism in Fig. 9.1. It is desired to determine the dimensions a, b, c and d that generate the desired function $\gamma = f(\theta)$.

---

[1] Ferdinand Freudenstein (1926–2006), a Hungarian-American physicist and engineer, is considered nowadays as the father of modern applied kinematics and a pioneer of the theory of machines and mechanisms in the United States.

D. Martins and E. H. Murai, *Mechanisms–Synthesis and Analysis with Applications in Robotics*, Mechanisms and Machine Science 183,
https://doi.org/10.1007/978-3-031-95187-9_9

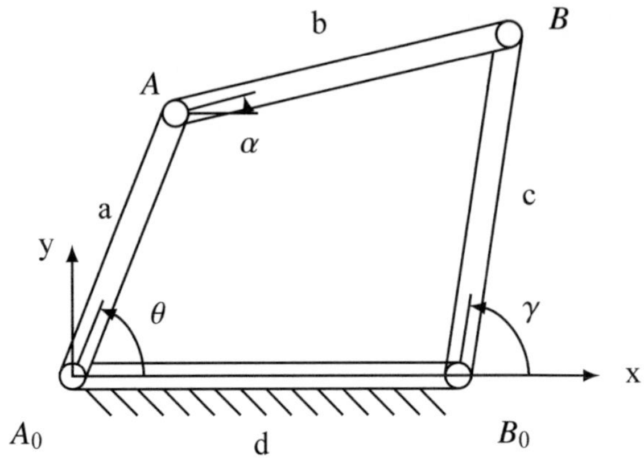

Fig. 9.1 Four-bar mechanism

Similarly to what was presented in Sect. 5.6.1, one can determine the Freudenstein equation from the geometry of the four-bar mechanism as being

$$\begin{aligned} x: &\quad a\cos(\theta) + b\cos(\alpha) - c\cos(\gamma) = d \\ y: &\quad a\sin(\theta) + b\sin(\alpha) = c\sin(\gamma). \end{aligned} \tag{9.1}$$

Isolating α from Eqs. 9.1, we get

$$\begin{aligned} x: &\quad b\cos(\alpha) = d - a\cos(\theta) + c\cos(\gamma) \\ y: &\quad b\sin(\alpha) = c\sin(\gamma) - a\sin(\theta). \end{aligned} \tag{9.2}$$

Squaring both Eqs. 9.2 and adding them together, we get

$$b^2(\cos^2(\alpha) + \sin^2(\alpha)) = (d - a\cos(\theta) + c\cos(\gamma))^2 + (c\sin(\gamma) - a\sin(\theta))^2. \tag{9.3}$$

Expanding the terms to the square and substituting $\cos^2(\alpha) + \sin^2(\alpha)$ for 1, we get

$$\begin{aligned} b^2 = d^2 + a^2\cos^2(\theta) + c^2\cos^2(\gamma) - 2ad\cos(\theta) + 2cd\cos(\gamma) \\ -2ac\cos(\theta)\cos(\gamma) + c^2\sin^2(\gamma) + a^2\sin^2(\theta) - 2ac\sin(\gamma)\sin(\theta). \end{aligned} \tag{9.4}$$

Rewritting the expression to factorate a^2, c^2 and $-2ac$,

$$\begin{aligned} b^2 = d^2 + a^2(\cos^2(\theta) + \sin^2(\theta)) + c^2(\cos^2(\gamma) + \sin^2(\gamma)) - 2ad\cos(\theta) \\ +2cd\cos(\gamma) - 2ac(\cos(\theta)\cos(\gamma) + \sin(\gamma)\sin(\theta)). \end{aligned} \tag{9.5}$$

By making a few trigonometric substitutions, we have

$$b^2 = d^2 + a^2 + c^2 - 2ad\cos(\theta) + 2cd\cos(\gamma) - 2ac\cos(\gamma - \theta). \tag{9.6}$$

Isolating $\cos(\gamma - \theta)$, we get

$$\cos(\gamma - \theta) = \frac{d^2 + a^2 + c^2 - b^2}{2ac} - \frac{d}{c}\cos(\theta) + \frac{d}{a}\cos(\gamma). \tag{9.7}$$

Equation 9.7 is known as the Freudenstein equation.

The function generation synthesis is a problem whose solution is scalable. Thus, given a solution with any values a, b, c and d, multiplying all the sizes by a **scale constant** k results in a new mechanism k times larger, but which still has the same function $\gamma = f(\theta)$. This shows that the result of the function generation is not just one solution, but a family of solutions.

By using four coordinate pairs, (θ, γ), Eq. 9.7 is written four times, and so one has a system with four equations and four unknowns (a, b, c, d).

To make it easier to solve, it is customary to consider one of the links with a given length, obtaining a system with three equations and three unknowns. Thus, by solving the function generation problem, the size of the other three links that satisfies $\gamma = f(\theta)$ is obtained. If a larger or smaller mechanism is required, simply multiply all dimensions by an appropriate scaling constant. To develop the function generation solution, this book adopts an unitary dimension for the fixed bar, $d = 1\, l.u.$ Substituting the dimension $d = 1\, l.u.$ and defining R_1, R_2 and R_3, we have

$$\begin{aligned} \cos(\gamma - \theta) &= R_1 + R_2\cos(\gamma) + R_3\cos(\theta) \\ R_1 &= \frac{a^2 - b^2 + c^2 + 1}{2ac} \\ R_2 &= +\frac{1}{a} \\ R_3 &= -\frac{1}{c}. \end{aligned} \tag{9.8}$$

Equations 9.8 have three unknowns, R_1, R_2 and R_3, so three equations are needed. By using three coordinate points $(\theta_i; \gamma_i)$ to describe the desired function $\gamma = f(\theta)$, one can write Eq. 9.8 three times.

This gives

$$\begin{aligned}\cos(\gamma_1 - \theta_1) &= R_1 + R_2 \cos(\gamma_1) + R_3 \cos(\theta_1)\\ \cos(\gamma_2 - \theta_2) &= R_1 + R_2 \cos(\gamma_2) + R_3 \cos(\theta_2)\\ \cos(\gamma_3 - \theta_3) &= R_1 + R_2 \cos(\gamma_3) + R_3 \cos(\theta_3).\end{aligned} \tag{9.9}$$

Note that R_1, R_2 and R_3 do not depend on θ or γ, so R_1, R_2 and R_3 are the same in the three Eqs. 9.9. Thus, we have a system of three equations and three unknowns. Once the values of R_1, R_2 and R_3 are determined, it is possible to calculate a, b and c using the respective Eqs. 9.8. The solution to Eqs. 9.9 can be written in matrix form as

$$\begin{bmatrix} R_1 \\ R_2 \\ R_3 \end{bmatrix} = \begin{bmatrix} 1 \ \cos(\gamma_1) \ \cos(\theta_1) \\ 1 \ \cos(\gamma_2) \ \cos(\theta_2) \\ 1 \ \cos(\gamma_3) \ \cos(\theta_3) \end{bmatrix}^{-1} \begin{bmatrix} \cos(\gamma_1 - \theta_1) \\ \cos(\gamma_2 - \theta_2) \\ \cos(\gamma_3 - \theta_3) \end{bmatrix}. \tag{9.10}$$

If a or c is negative, the link must be oriented in the opposite direction. Thus, the direction of links $\overline{A_0A}$ and $\overline{B_0B}$ is given by θ and γ, respectively, but the sense is given by the sign, where for positive values of θ and γ, it is used θ and γ; and for negative values of θ and γ, it is used $\theta + 180°$ and $\gamma + 180°$.

Summary of the method:

1. Given three coordinate pairs $(\theta; \gamma)$, substitute in Eq. 9.10 to determine R_1, R_2 and R_3.
2. Substitute R_2 and R_3 into the last two Eqs. 9.8 to determine a and c, respectively.
3. Substitute R_1, a, and c into the second Eq. 9.8 to determine b.
4. If necessary, multiply a, b, c and d by a scaling factor.

Function generation

Given the three coordinate pairs $(\theta; \gamma)$ in Table 9.1, determine the dimensions of the four-bar mechanism so that it satisfies this requirement.

Table 9.1 Coordinate points for the desired $\gamma = f(\theta)$ function

θ	40°	70°	75°
γ	100°	115°	120°

Solution:

Substituting the values of θ and γ in Eq. 9.10, we get

$$\begin{bmatrix} R_1 \\ R_2 \\ R_3 \end{bmatrix} = \begin{bmatrix} 1 \cos(100^\circ) \cos(40^\circ) \\ 1 \cos(115^\circ) \cos(70^\circ) \\ 1 \cos(120^\circ) \cos(75^\circ) \end{bmatrix}^{-1} \begin{bmatrix} \cos(100^\circ - 40^\circ) \\ \cos(115^\circ - 70^\circ) \\ \cos(120^\circ - 75^\circ) \end{bmatrix}. \quad (9.11)$$

Solving the system of Eqs. 9.11, the solution using four decimal places is

$$\begin{aligned} R_1 &= 1.7622 \\ R_2 &= 1.4244 \\ R_3 &= -1.3248. \end{aligned} \quad (9.12)$$

Setting up the second system of equations to determine a, b and c

$$\begin{aligned} 1.7622 &= \frac{a^2 - b^2 + c^2 + 1}{2ac} \\ 1.4244 &= +\frac{1}{a} \\ -1.3248 &= -\frac{1}{c}. \end{aligned} \quad (9.13)$$

Solving this second system of equations, we get

$$\begin{aligned} b &= 0.4416\, l.u. \\ a &= 0.7020\, l.u. \\ c &= 0.7548\, l.u.. \end{aligned} \quad (9.14)$$

If it is desired that the fixed bar has a length of 3 $l.u.$, simply multiply the size of all bars by 3, giving:

$$\begin{aligned} a &= 2.1061\, l.u. \\ b &= 1.3247\, l.u. \\ c &= 2.2645\, l.u. \\ d &= 3\, l.u. \end{aligned} \quad (9.15)$$

Figure 9.2 shows the four-bar mechanism designed using function generation.

More content regarding function generation can be seen at [2–4].

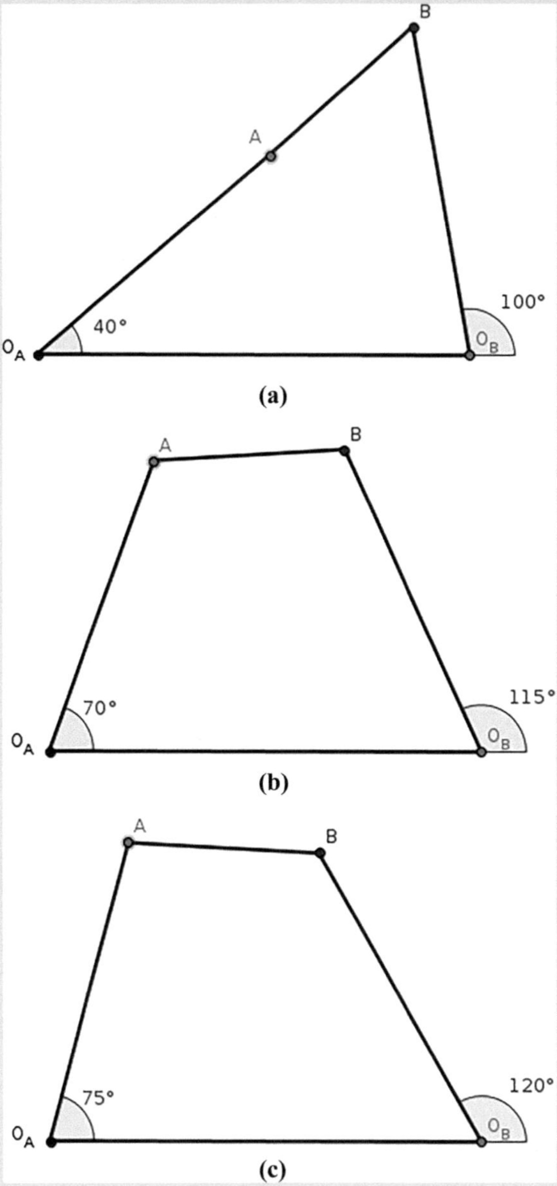

Fig. 9.2 Four-bar mechanism designed using function generation. **a** $\theta = 40°$. **b** $\theta = 70°$. **c** $\theta = 75°$

9.2 Crank-rocker Dimensional Synthesis

The aim of the crank-rocker dimensional synthesis is to determine the dimensions of a four-bar mechanism with a crank input and a rocker output at a certain specified angle. This synthesis is useful when it is desired an oscillatory motion with amplitude $\Delta\gamma$ using a motor as an actuator (crank input).

This synthesis is used as a complement to 2 PFS and 3 PFS syntheses, since these syntheses generate mechanisms that are used as double-rockers. Thus, one can use 2 PFS and 3 PFS syntheses to obtain a primary four-bar mechanism that reaches the desired positions and then use the crank-rocker synthesis to obtain a second four-bar mechanism that will drive the first four-bar mechanism. Note that by connecting these two four-bar mechanisms, a Watt mechanism is obtained.

There are two methods for making the crank-rocker synthesis: the geometric method and the algebraic method, which will be seen in Sects. 9.2.1 and 9.2.2, respectively.

9.2.1 Geometric Method

This section presents the dimensional synthesis of the crank-rocker using the geometric method. Initially, the position for the fixed pivot of the output bar, B_0, is chosen. Then choose the size of the output bar (c) and draw a circle centered on B_0 and with radius c, as shown in Fig. 9.3.

Two points are then chosen on the circumference that represent the rocker limit positions, B_1 and B_2, as shown in Fig. 9.4. Draw the line through B_1 and B_2. Define the line segment l to be $\overline{B_1 B_2}$, see Fig. 9.5. On the line $\overline{B_1 B_2}$ choose the position of A_0, as indicated by the Fig. 9.6. Next, draw a circle centered on A_0 and of radius

Fig. 9.3 Choice of B_0 and c

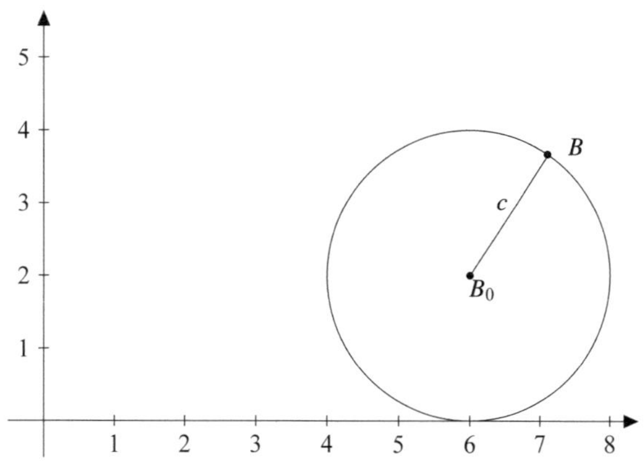

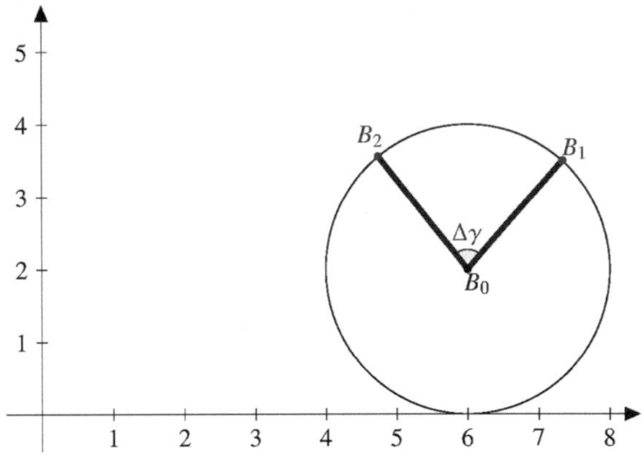

Fig. 9.4 Choice of B_1 and B_2

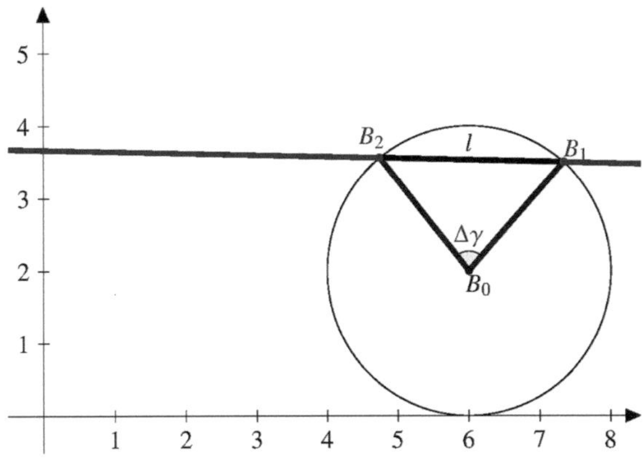

Fig. 9.5 Line $\overline{B_1 B_2}$ and line segment l

$l/2$, as shown in Fig. 9.6. The radius of the circle is the size of the input bar, and the points of intersection of the circle with the line $\overline{B_1 B_2}$ are the points A_1 and A_2.

When the crank-rocker is in its furthest position (to the right), the input bar moving pivot is in position A_1. When the crank-rocker is in its closest position (more to the left), the input bar moving pivot is in position A_2. The crank-rocker mechanism can thus be drawn as:

Input bar: $\overline{A_0 A_1}$.
Coupler bar: $\overline{A_1 B_1}$.
Output bar: $\overline{B_0 B_1}$.
Fixed bar: $\overline{A_0 B_0}$.

The mechanism in an intermediary position is shown in Fig. 9.7. The mechanism from Fig. 9.6 is shown again in Fig. 9.8 with the kinematic pairs aligned but without overlapping links, illustrating how each crank-rocker link would be arranged in the extreme positions of the crank-rocker.

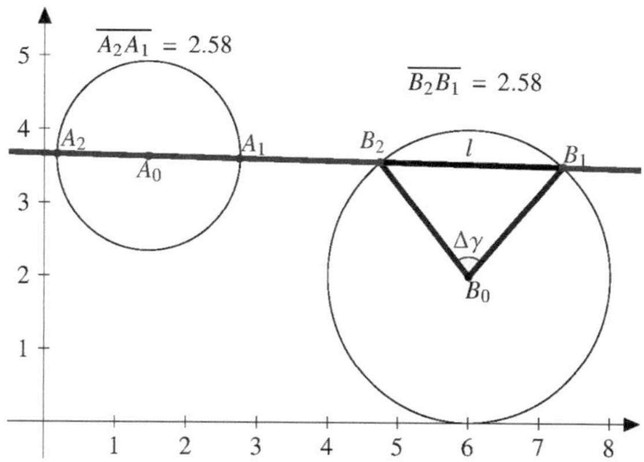

Fig. 9.6 Choice of A_0

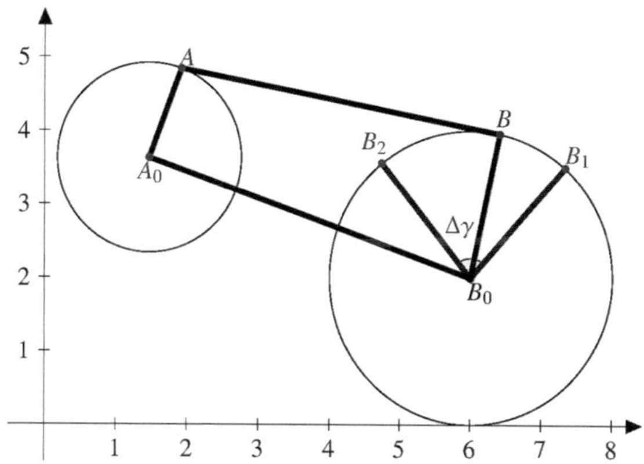

Fig. 9.7 Crank-rocker mechanism

Summary of the method:

1. Choose the position of B_0.
2. Choose the size of the output bar, c.
3. Draw a circle centered on B_0 and of radius c.
4. Mark the two limit positions of the rocker, B_1 and B_2.
5. Draw the line $\overline{B_1B_2}$ and measure the length of the line segment l.
6. Choose the position of A_0 on the line $\overline{B_1B_2}$.
7. Draw a circle centered at A_0 and with radius $l/2$.
8. Determine the crank-rocker with points A_0, A_1, B_0 and B_1.

9.2.2 Algebraic Method

Consider the four-bar mechanism with the rocker limited to the two positions as shown in Fig. 9.9.

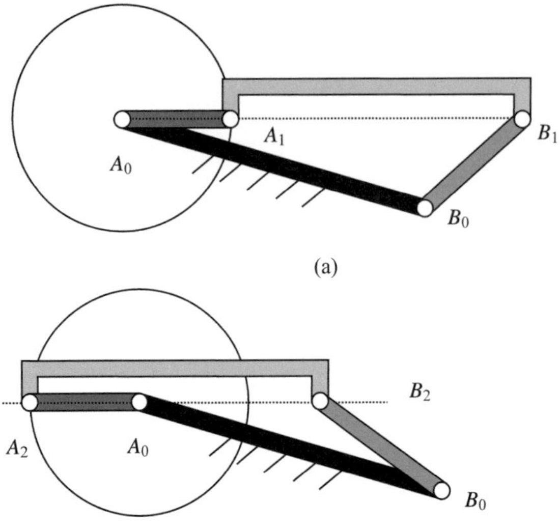

Fig. 9.8 Crank-rocker linkage with crank and coupler links aligned

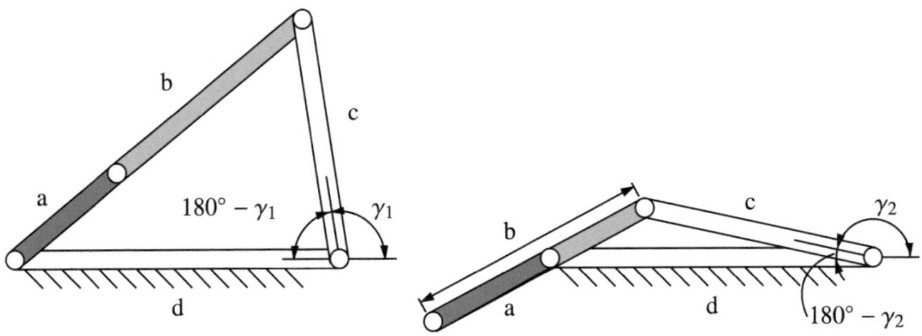

Fig. 9.9 Crank-rocker limit positions

Note that the positions shown in Fig. 9.9 are the limit positions, since in the case of Fig. 9.9a the input link cannot move the output link in such a way as to reduce γ_1. Similarly, in Fig. 9.9b the input link cannot move the output link in such a way as to increase γ_2. Considering the inner triangle in Figs. 9.9a, b, using the law of cosines results in

$$\begin{aligned} &(a) : (b+a)^2 = c^2 + d^2 - 2cd\cos(180^\circ - \gamma_1) \\ &(b) : (b-a)^2 = c^2 + d^2 - 2cd\cos(180^\circ - \gamma_2). \end{aligned} \tag{9.16}$$

Adding Eqs. 9.16(a) and 9.16(b) and dividing by two, we get

$$a^2 + b^2 = c^2 + d^2 - cd(\cos(180° - \gamma_1) + \cos(180° - \gamma_2)). \tag{9.17}$$

Subtracting Eq. 9.16(b) from Eq. 9.16(a) and dividing by two, we get

$$2ab = cd(\cos(180° - \gamma_2) - \cos(180° - \gamma_1)). \tag{9.18}$$

Isolating b from Eq. 9.18, we get

$$b = \frac{cd}{2a}\left(\cos(180° - \gamma_2) - \cos(180° - \gamma_1)\right). \tag{9.19}$$

Substituting b into Eq. 9.17, we get

$$\begin{aligned} a^2 + \left(\frac{cd}{2a}(\cos(180° - \gamma_2) - \cos(180° - \gamma_1))\right)^2 &= c^2 + d^2 \\ -cd(\cos(180° - \gamma_1) + \cos(180° - \gamma_2)). \end{aligned} \tag{9.20}$$

Rearranging the terms of Eq. 9.20, factoring the terms with $\boldsymbol{c^2}$ and $\boldsymbol{c}$, obtaining

$$\begin{aligned} c^2\left(1 - \frac{d^2}{4a^2}(\cos(180° - \gamma_2) - \cos(180° - \gamma_1))^2\right) - cd(\cos(180° - \gamma_1) \\ + \cos(180° - \gamma_2)) + d^2 - a^2 = 0. \end{aligned} \tag{9.21}$$

Defining P, Q and R and rewriting Eq. 9.21

$$\begin{aligned} Pc^2 + Qc + R = 0. \\ P = 1 - \frac{d^2}{4a^2}(\cos(180° - \gamma_2) - \cos(180° - \gamma_1))^2 \\ Q = -d\,(\cos(180° - \gamma_1) + \cos(180° - \gamma_2)) \\ R = d^2 - a^2 \end{aligned} \tag{9.22}$$

Note that P, Q and R depend only on a, d, γ_1 and γ_2. The angle of oscillation ($\Delta\gamma$) is known, and you can choose an initial angle γ_1 and determine γ_2. The dimensions a and d must be chosen. You can then determine the values of P, Q and R by substituting a, d, γ_1 and γ_2 into Eqs. 9.22.

Once you have determined the coefficients P, Q and R, you can solve the second-degree equation whose variable is c, shown in Eq. 9.22. Finally, the value of b can be calculated using Eq. 9.19.

Note that negative values for the size of the links should be discarded. Furthermore, given the sizes a, b, c and d, the four-bar mechanism can be assembled in two different working modes. One working mode will have an oscillation in the output bar from γ_1 to γ_2 while the other working mode will have an oscillation from $-\gamma_1$ to $-\gamma_2$.

Summary of the method:

1. Choose appropriate values for γ_1, a and d.
2. Given a $\Delta\gamma$, determine γ_2.
3. Determine P, Q and R using the last three Eqs. 9.22.
4. Substitute P, Q and R into the first Eq. 9.22 and solve the quadratic equation to determine c values.
5. Substitute a, c, d, γ_1 and γ_2 into Eq. 9.19 to determine the value of b.
6. If necessary, multiply a, b, c and d by a scale factor.

Analytical crank-rocker synthesis

In Example 1 it was designed a windshield wiper using a double-rocker four-bar mechanism and a motor to drive the mechanism. In this example we want to design a crank-rocker mechanism to drive the wiper. As a specification, the angle of oscillation should be $\Delta\gamma = 117°$ (same angle as in Example 1).

Solution:

First determine the values for a, d, γ_1. To do this, make an initial sketch of what the mechanism will look like. Starting with the double-rocker four-bar mechanism in Fig. 9.10a, highlighted in gray, we insert the crank-rocker mechanism, as shown in Fig. 9.10b. The motor and the space that the crank-rocker mechanism occupies when in operation must be below the car hood, hence the arrangement adopted in Fig. 9.10b for the motor and pivots.

The angle γ is measured from the ray $\overrightarrow{A_0B_0}$ to the ray $\overrightarrow{B_0B}$. There are two ways to measure this angle, clockwise and counterclockwise. To avoid measuring in the wrong direction, one can analyze which of the angles between the two aforementioned rays is supplementary to the angle of the triangle. To make it clearer, just take a look at Fig. 9.9. One can see that in the limit positions the mechanism forms a triangle, and the angle γ is measured so that it is supplementary to the respective angle of the triangle. The angle γ must be measured in this way because the formulation was developed using this angle. Figure 9.10c shows how γ should be measured.

Having established how γ should be measured, γ_1 is determined. In the initial condition, the windshield wiper is at rest, as shown in Fig. 9.10d. As this is a boundary condition, the input and coupler links of the crank-rocker must be aligned as shown in Fig. 9.10d. For the situation in Fig. 9.10d, measuring the angle yields $\gamma_1 = 35°$. Analyzing the dimensions of Fig. 9.10d, an initial estimate of $a = 90$ mm and $d = 300$ mm can be done. The estimated values of γ_1, a and d were based on the positioning of the motor in relation to the existing double rocker mechanism. Therefore, a good motor positioning is essential for a good initial estimate of a, d and γ_1.

The angle γ_2 is calculated by

$$\gamma_2 = \gamma_1 + \Delta\gamma = 35^\circ + 117^\circ = 152^\circ \tag{9.23}$$

Substituting γ_1, γ_2, a and d into the equations for P, Q and R, we get

$$\begin{aligned} P &= 1 - \frac{0.3^2}{4 * 0.09^2}(\cos(180^\circ - 152^\circ) - \cos(180^\circ - 35^\circ))^2 = -7.04762 \\ Q &= -0.3\,(\cos(180^\circ - 35^\circ) + \cos(180^\circ - 152^\circ)) = -0.01914\,m \\ R &= 0.3^2 - 0.09^2 = 0.08190\,m^2. \end{aligned} \tag{9.24}$$

Substituting P, Q and R into Eq. 9.22, we get

$$-7.04762c^2 - 0.01914c + 0.08190 = 0. \tag{9.25}$$

Solving the second-degree equation gives the values for c

$$c_1 = -0.10916\,m \qquad c_2 = 0.10645\,m. \tag{9.26}$$

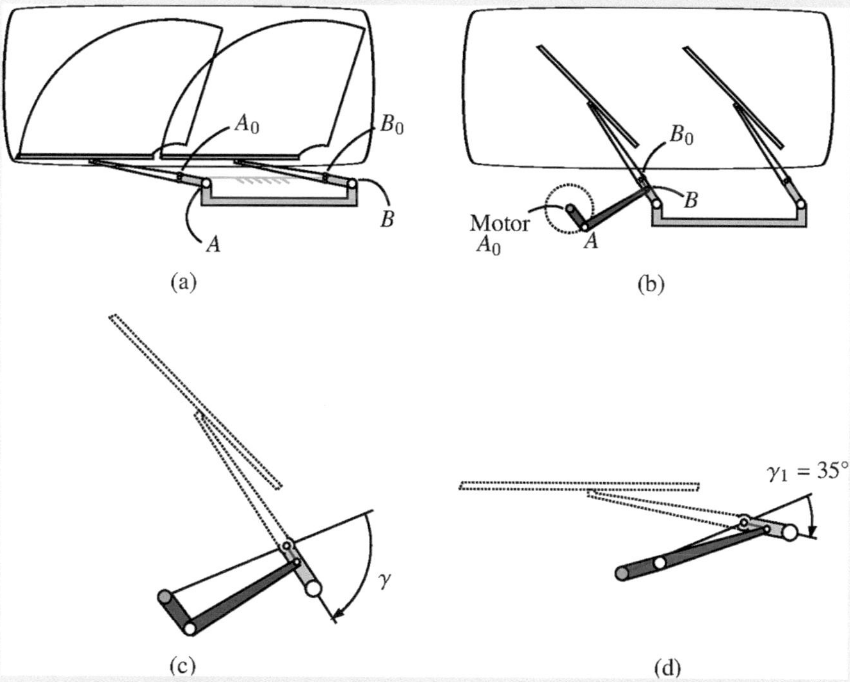

Fig. 9.10 Determining the angle γ_1

Thus, the dimension of the coupler bar is $c = 106.45$ mm. Substituting a, c, d, γ_1 and γ_2 into Eq. 9.19, we get

$$b = \frac{0.10645 * 0.3}{2 * 0{,}09}(\cos(180^\circ - 152^\circ) - \cos(180^\circ - 35^\circ)) = 0.30198\ m \tag{9.27}$$

Thus, rounding the dimensions to the nearest millimeter, we have $a = 90$ mm, $b = 302$ mm, $c = 106$ mm and $d = 300$ mm. These dimensions are feasible for the design of a windshield wiper drive, since the space occupied by this mechanism when operating corresponds to what was initially planned when positioning the motor.

If necessary, the entire mechanism could be enlarged or reduced using a scale factor. Also, if the angle γ_1 needed to be changed, this could be done by changing the position of the pivot B of the crank-rocker, as shown in Fig. 9.11.

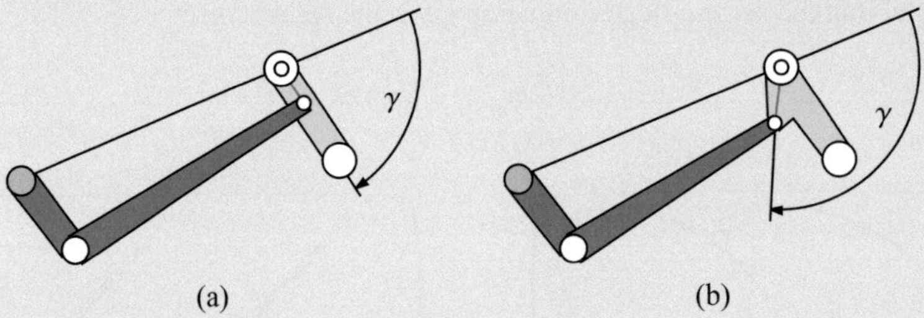

Fig. 9.11 γ_1 angle adjustment

Problems

9.1 Design a mechanism that generates a function described by the points in Table 9.2. Synthesize a mechanism with:

(a) $d = 5\ l.u.$
(b) $a = 3\ l.u.$
(c) $c = 2\ l.u.$

9.2 Using the geometric method, determine a crank-rocker with the following characteristics:

- $\Delta\gamma = 30^\circ$
- $d = 4\ l.u.$
- $a = 0.5\ l.u.$

Table 9.2 Points for function generation

θ	40°	60°	70°
γ	90°	110°	125°

9.3 Using the algebraic method, determine a crank-rocker with the following characteristics:

- $\gamma_1 = 50^\circ$
- $\gamma_2 = 80^\circ$
- $d = 4\,l.u.$
- $a = 0.5\,l.u.$

9.4 Using the geometric method, determine a crank-rocker with the following characteristics:

- $\Delta\gamma = 90^\circ$
- $d = 10\,l.u.$
- $c = 4\,l.u.$

9.5 Given the four dimensions of a crank-rocker, $a = 2\,l.u.$, $b = 5.5\,l.u.$, $c = 4\,l.u.$ and $d = 6\,l.u.$, determine γ_1 and γ_2.

References

1. Freudenstein, F.: Approximate synthesis of four-bar linkages. Trans. ASME **77**(6), 853–859 (1955)
2. Suh, C., Radcliffe, C.: Kinematics and Mechanisms Design. R.E. Krieger Pub. Co., Malabar (1978)
3. Hartenberg, R.S., Denavit, J.: Kinematic Synthesis of Linkages. McGraw-Hill, New York (1964)
4. Hunt, K.H.: Kinematic Geometry of Mechanisms. Clarendon Press, Oxford (1978)

Appendix A
Script for the Transmission Quality Graph

```
%%%%%%%%%%%%%%%%%%%%%%%%%%%%%%%%%%%%%%%%%%%%%%%%%%%%%%%%%%%
%
% SCRIPT FOR DETERMINING THE TRANSMISSION ANGLE GRAPH
% IMPLEMENTED IN OCTAVE
%
%%%%%%%%%%%%%%%%%%%%%%%%%%%%%%%%%%%%%%%%%%%%%%%%%%%%%%%%%%%
close all
clear all
clc
%%%%%%%%%%%%%%%%%%%%%%%%%%%%%%%%%%%%%%%%%%%%%%%%%%%%%%%%%%%
% INPUT DATA
%%%%%%%%%%%%%%%%%%%%%%%%%%%%%%%%%%%%%%%%%%%%%%%%%%%%%%%%%%%

a=4;
b=2;
c=4;
d=5;

%%%%%%%%%%%%%%%%%%%%%%%%%%%%%%%%%%%%%%%%%%%%%%%%%%%%%%%%%%%
% GRASHOF
%%%%%%%%%%%%%%%%%%%%%%%%%%%%%%%%%%%%%%%%%%%%%%%%%%%%%%%%%%%

v=[a, b, c, d];
if max(v)+min(v) <= sum(v)-max(v)-min(v)
    disp('GRASHOF')
    grashof=1;
else
    disp('NOT GRASHOF')
    grashof=0;
end
```


D. Martins and E. H. Murai, *Mechanisms–Synthesis and Analysis with Applications in Robotics*, Mechanisms and Machine Science 183,
https://doi.org/10.1007/978-3-031-95187-9

```
%%%%%%%%%%%%%%%%%%%%%%%%%%%%%%%%%%%%%%%%%%%%%%%%%%%%%%%%%%%%%%
% ROCKER LIMITS WHEN THE INPUT IS THE ROCKER
%%%%%%%%%%%%%%%%%%%%%%%%%%%%%%%%%%%%%%%%%%%%%%%%%%%%%%%%%%%%%%

if grashof==0 || (grashof==1&&b==min(v)) % If it is a rocker
    x1 = (a^2-(b+c)^2+d^2)/(2*d);
    x2 = (a^2-(b-c)^2+d^2)/(2*d);
    teta1 = acosd(x1/a);
    teta2 = acosd(x2/a);
    A = (b^2+c^2-a^2-d^2)/(2*b*c);
    B = (a*d)/(b*c);
    theta = teta2:1:teta1;
else % if it is not a rocker
    A = (b^2+c^2-a^2-d^2)/(2*b*c);
    B = (a*d)/(b*c);
    theta = 0:1:360;
end
mu = acosd(A+B*cosd(theta));
figure()
plot(theta, mu)
grid on
title('Transmission angle')
xlabel('Input angle')
ylabel('Transmission angle')
hold on
plot([theta(1),theta(size(theta, 2))], [140,140], '-r')
plot([theta(1),theta(size(theta, 2))], [40,40], '-r')
```

Appendix B
Script for the Slider-Crank Kinematic Analysis

```
%%%%%%%%%%%%%%%%%%%%%%%%%%%%%%%%%%%%%%%%%%%%%%%%%%%%%%%%%%%%%
%
% SLIDER-CRANK POSITION ANALYSIS IMPLEMENTED IN OCTAVE
%
%%%%%%%%%%%%%%%%%%%%%%%%%%%%%%%%%%%%%%%%%%%%%%%%%%%%%%%%%%%%%
close all
clear all
clc
%%%%%%%%%%%%%%%%%%%%%%%%%%%%%%%%%%%%%%%%%%%%%%%%%%%%%%%%%%%%%
% INPUT DATA
%%%%%%%%%%%%%%%%%%%%%%%%%%%%%%%%%%%%%%%%%%%%%%%%%%%%%%%%%%%%%
a = 2;
b = 4;
d_1 = 0.5;
dteta = 1; %rad/s
ddteta = 0; % rad/s^2

%%%%%%%%%%%%%%%%%%%%%%%%%%%%%%%%%%%%%%%%%%%%%%%%%%%%%%%%%%%%%
% POSITION GRAPH
%%%%%%%%%%%%%%%%%%%%%%%%%%%%%%%%%%%%%%%%%%%%%%%%%%%%%%%%%%%%%
teta = [0:1:360];
beta=asind((a*sind(teta)-d_1)/(b));
x=a*cosd(teta)+b*cosd(beta);

figure()
plot(teta, beta, '-r')
grid on
title('Position analysis')
xlabel('Input angle')
ylabel('Beta angle')
```


D. Martins and E. H. Murai, *Mechanisms–Synthesis and Analysis with Applications in Robotics*, Mechanisms and Machine Science 183,
https://doi.org/10.1007/978-3-031-95187-9

```
figure()
plot(teta, x, '-r')
grid on
title('Position analysis')
xlabel('Input angle')
ylabel('Distance x')

%%%%%%%%%%%%%%%%%%%%%%%%%%%%%%%%%%%%%%%%%%%%%%%%%%%%%%%%%%%%
% VELOCITY GRAPH
%%%%%%%%%%%%%%%%%%%%%%%%%%%%%%%%%%%%%%%%%%%%%%%%%%%%%%%%%%%%

dbeta = (a*cosd(teta)*dteta)./(b*cosd(beta));
dx=-a*sind(teta).*dteta-b*sind(beta).*dbeta;

figure()
plot(teta, dbeta, '-r')
grid on
title('Velocity analysis')
xlabel('Input angle')
ylabel('Beta velocity')

figure()
plot(teta, dx, '-r')
grid on
title('Velocity analysis')
xlabel('Input angle')
ylabel('Velocity x')

%%%%%%%%%%%%%%%%%%%%%%%%%%%%%%%%%%%%%%%%%%%%%%%%%%%%%%%%%%%%
% ACCELERATION GRAPH
%%%%%%%%%%%%%%%%%%%%%%%%%%%%%%%%%%%%%%%%%%%%%%%%%%%%%%%%%%%%

ddbeta=(-a*sind(teta).*dteta.^2 + a*cosd(teta).*ddteta +
b*sind(beta).*dbeta.^2)./(b*cosd(beta));
ddx=-a*cosd(teta).*dteta.^2-a*sind(teta).*ddteta-
b*cosd(beta).*dbeta.^2-b*sind(beta).*ddbeta;

figure()
plot(teta, ddbeta, '-r')
grid on
title('Acceleration analysis')
xlabel('Input angle')
ylabel('Beta acceleration')
```

```
figure()
plot(teta, ddx, '-r')
grid on
title('Acceleration analysis')
xlabel('Input angle')
ylabel('Acceleration x')
```

Solutions

Problems of Chapter 1

1.1 Similar to Example 1, there is more than one solution to the graph. The mechanism has eight links of which four are ternary (links 1, 2, 5, and 6) and four are binary (links 3, 4, 7, and 8). The types of links can vary depending on the generated graph. There are 10 kinematic pairs (see the number of edges in the graph). All kinematic pairs are revolute. The mechanism has three independent loops (Fig. 1).

1.2 The mechanism has six links being two ternary links (links 3 and 6) and four binary links (links 1, 2, 4 and 5). There are seven kinematic pairs (see the number of edges in the graph). All kinematic pairs are revolute. The mechanism has two independent loops (Watt's kinematic chain) (Fig. 2).

1.3 The mechanism has four links, all of which are binary. There are four kinematic pairs (see the number of edges in the graph). The kinematic pair connecting links 1 and 4 is prismatic, the others are revolute. The mechanism has an independent loop (Fig. 3).

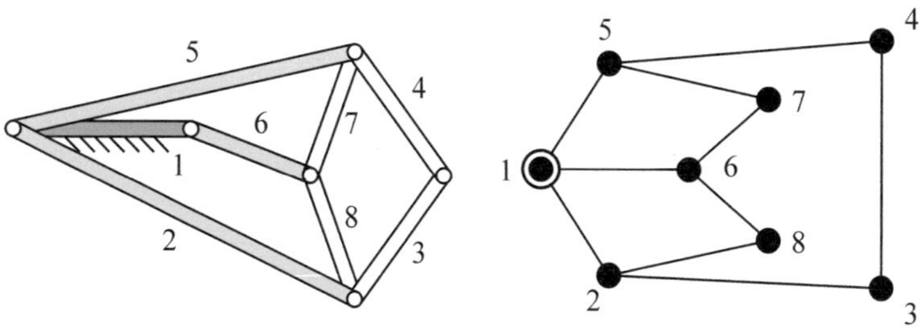

Fig. 1 Peaucellier mechanism

D. Martins and E. H. Murai, *Mechanisms–Synthesis and Analysis with Applications in Robotics*, Mechanisms and Machine Science 183,
https://doi.org/10.1007/978-3-031-95187-9

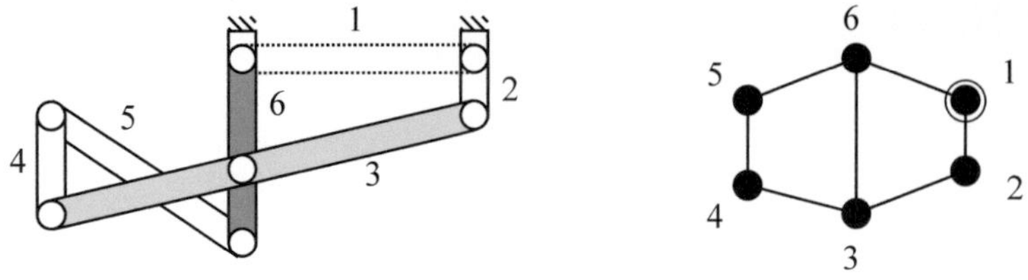

Fig. 2 *Flap* mechanisms

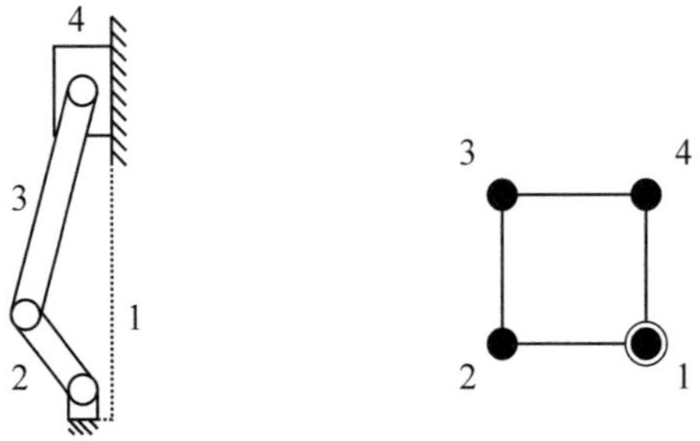

Fig. 3 Slider-crank mechanism

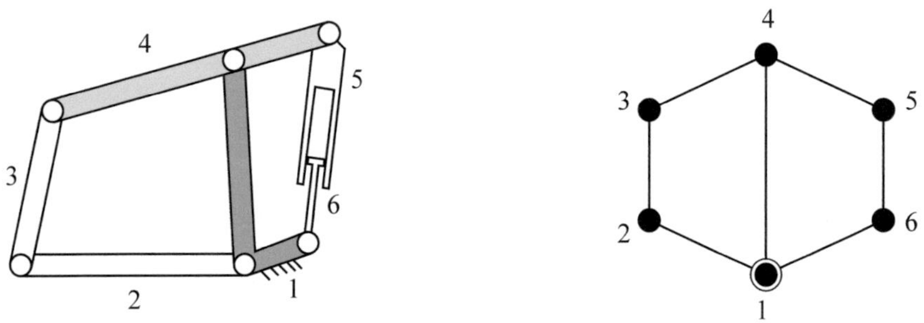

Fig. 4 Suspension mechanism

1.4 The mechanism has six links being two ternary links (links 1 and 4) and four binary links (links 2, 3, 5 and 6). There are seven kinematic pairs (see the number of edges in the graph). The kinematic pair connecting links 5 and 6 is prismatic, the others are revolute. The mechanism has two independent loops (Watt's kinematic chain) (Fig. 4).

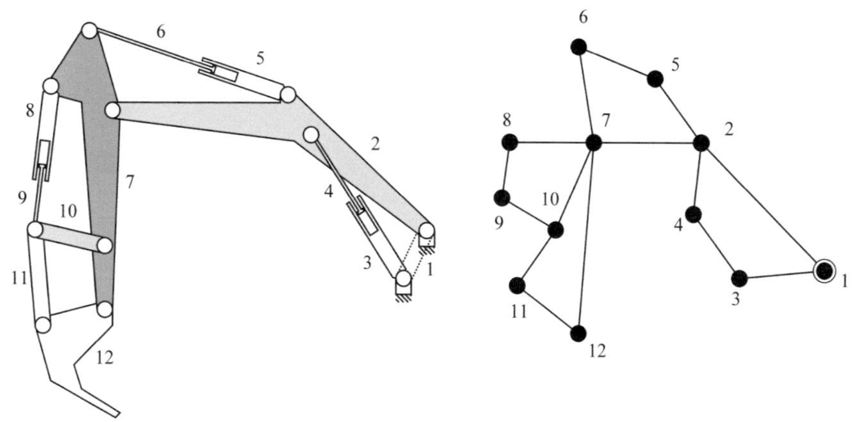

Fig. 5 Mechanism for excavator arm

1.5 Similar to Example 1, there is more than one solution for the graph. The mechanism has 12 links being one quinary (link 7), one quaternary (link 2), one ternary (link 10) and nine binary links (links 1, 3, 4, 5, 6, 8, 9, 11 and 12). The types of links can vary depending on the generated graph. There are 15 kinematic pairs (see the number of edges in the graph). The kinematic pairs connecting links 3 and 4; 5 and 6; 8 and 9 are prismatic pairs, the rest are revolute pairs. The mechanism has four independent loops (Fig. 5).

Problems of Chapter 2

2.1 $M = 1$
2.2 $M = 1$
2.3 $M = 1$
2.4 $M = 1$
2.5 $M = 3$
2.6 $M = 2$
2.7 $M = 2$

Problems of Chapter 3

3.1 $\mu_{min} = 33.56°$ and $\mu_{max} = 72.54°$
3.2 $a = 0.4926\ l.u.$, $b = 5\ l.u.$, $c = 1\ l.u.$ and $d = 5.0752\ l.u.$
3.3 $\mu_{min} \approx 51.7°$. $\mu_{max} \approx 138.4°$. The transmission angle is within the recommended range.

Problems of Chapter 4

4.1 $R = \begin{bmatrix} 0&0&0&0&1&0&0&0 \\ 0&0&0&0&0&1&0&0 \\ 0&0&0&0&0&0&1&0 \\ 0&0&0&0&0&0&0&1 \\ 1&0&0&0&0&0&0&0 \\ 0&1&0&0&0&0&0&0 \\ 0&0&1&0&0&0&0&0 \\ 0&0&0&1&0&0&0&0 \end{bmatrix}$

4.2 $V = M - 1$

4.3

(a) $M = 3$

(b) The connectivity and degree of control matrices will be identical.

(c) The redundancy matrix will be null.

(d) $C = K = \begin{bmatrix} 0&1&1&1&1&1&1&1&2&3&2&1 \\ 1&0&1&1&1&1&1&2&3&3&2&1 \\ 1&1&0&1&1&1&1&2&3&2&1&1 \\ 1&1&1&0&1&1&1&2&3&3&2&1 \\ 1&1&1&1&0&1&1&2&3&3&2&1 \\ 1&1&1&1&1&0&1&2&3&3&2&1 \\ 1&1&1&1&1&1&0&2&3&3&2&1 \\ 1&2&2&2&2&2&2&0&1&2&3&2 \\ 2&3&3&3&3&3&3&1&0&1&2&3 \\ 3&3&2&3&3&3&3&2&1&0&1&3 \\ 2&2&1&2&2&2&2&3&2&1&0&2 \\ 1&1&1&1&1&1&1&2&3&3&2&0 \end{bmatrix}$ $R = 0_{12 \times 12}$

Note: links 1, 2, 3, 4, 5, 6, 7 and 12 form a subchain with mobility 1.

(e) $V = 2$

4.4

(a) $K = C = \begin{bmatrix} 0&1&1&1&1&1 \\ 1&0&1&1&1&1 \\ 1&1&0&1&1&1 \\ 1&1&1&0&1&1 \\ 1&1&1&1&0&1 \\ 1&1&1&1&1&0 \end{bmatrix}$ $R = 0_{6 \times 6}$ $V = 0$

(b) $K = C = \begin{bmatrix} 0&1&2&3&2&1&1&2&1&1 \\ 1&0&1&2&3&2&2&2&1&1 \\ 2&1&0&1&2&3&2&1&2&2 \\ 3&2&1&0&1&2&3&2&3&3 \\ 2&3&2&1&0&1&3&3&3&3 \\ 1&2&3&2&1&0&2&3&2&2 \\ 1&2&2&3&3&2&0&1&2&2 \\ 2&2&1&2&3&3&1&0&3&3 \\ 1&1&2&3&3&2&2&3&0&1 \\ 1&1&2&3&3&2&2&3&1&0 \end{bmatrix} R = 0_{10 \times 10} \; V = 2$

(c) $K = C = \begin{bmatrix} 0&1&1&1&1&1&1&1 \\ 1&0&1&1&1&1&1&1 \\ 1&1&0&1&1&1&1&1 \\ 1&1&1&0&1&1&1&1 \\ 1&1&1&1&0&1&1&1 \\ 1&1&1&1&1&0&1&1 \\ 1&1&1&1&1&1&0&1 \\ 1&1&1&1&1&1&1&0 \end{bmatrix} R = 0_{8 \times 8} \; V = 0$

(d) likewise item c

(e) $K = C = \begin{bmatrix} 0&1&2&3&2&1 \\ 1&0&1&2&3&2 \\ 2&1&0&1&2&3 \\ 3&2&1&0&1&2 \\ 2&3&2&1&0&1 \\ 1&2&3&2&1&0 \end{bmatrix} R = 0_{6 \times 6} \; V = 0$

(f) $K = C = \begin{bmatrix} 0&1&1&1&1&1&1&1 \\ 1&0&1&1&1&1&1&1 \\ 1&1&0&1&1&1&1&1 \\ 1&1&1&0&1&1&1&1 \\ 1&1&1&1&0&1&1&1 \\ 1&1&1&1&1&0&1&1 \\ 1&1&1&1&1&1&0&1 \\ 1&1&1&1&1&1&1&0 \end{bmatrix} R = 0_{8 \times 8} \; V = 0$

Problems of Chapter 6

6.1 Solution of: $M = 3$, $\nu = 2$, $\lambda = 3$ (planar) (Figs. 6 and 7).

6.2 It cannot be solved because $\nu = 1.5$.

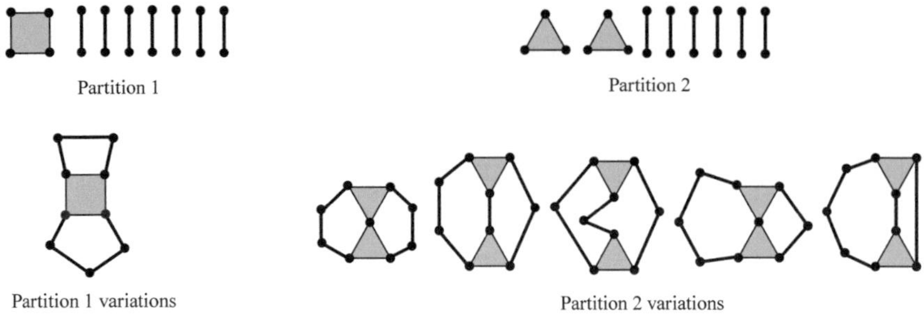

Fig. 6 Partitions, variations and inversions for $M = 3$, $\nu = 2$, $\lambda = 3$

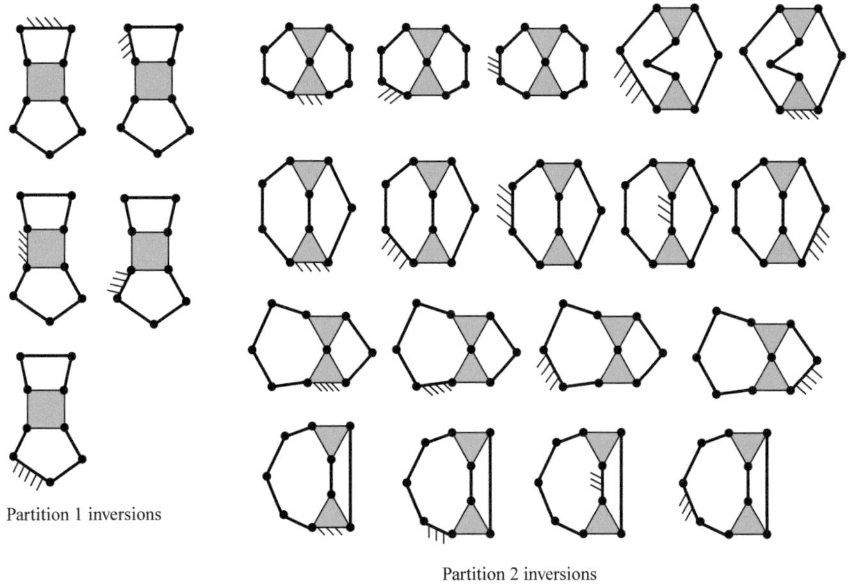

Fig. 7 Partitions, variations and inversions for $M = 3$, $\nu = 2$, $\lambda = 3$ (continued)

Problems of Chapter 9

9.1

a) $a = 2.056\ l.u.$; $b = 3.436\ l.u.$; $c = 1.599\ l.u.$; $d = 5\ l.u.$
b) $a = 3\ l.u.$; $b = 5.014\ l.u.$; $c = 2.334\ l.u.$; $d = 7.295\ l.u.$
c) $a = 2.571\ l.u.$; $b = 4.297\ l.u.$; $c = 2\ l.u.$; $d = 6.252\ l.u.$

9.3 $a = 0.5\ l.u.$; $b = 6.060\ l.u.$; $c = 3.229\ l.u.$; $d = 4\ l.u.$
9.5 $\gamma_1 = 84.920°$; $\gamma_2 = 145.907°$

Index

D. Martins and E. H. Murai, *Mechanisms–Synthesis and Analysis with Applications in Robotics*, Mechanisms and Machine Science 183,
https://doi.org/10.1007/978-3-031-95187-9

MIX
Papier aus verantwortungsvollen Quellen
Paper from responsible sources
FSC® C105338

If you have any concerns about our products,
you can contact us on
ProductSafety@springernature.com

In case Publisher is established outside the EU,
the EU authorized representative is:
Springer Nature Customer Service Center GmbH
Europaplatz 3, 69115 Heidelberg, Germany

Printed by Libri Plureos GmbH
in Hamburg, Germany